Novel Aspects of Gas Chromatography and Chemometrics

Edited by Serban C. Moldoveanu, Victor David and Vu Dang Hoang

Published in London, United Kingdom

Novel Aspects of Gas Chromatography and Chemometrics
http://dx.doi.org/10.5772/intechopen.102270
Edited by Serban C. Moldoveanu, Victor David and Vu Dang Hoang

Contributors
Nam-Ky Nguyen, Tung-Dinh Pham, Mai Phuong Vuong, Stella Stylianou, Biljana Otašević, Jovana Krmar, Nevena Đajić, Bojana Svrkota, Jevrem Stojanović, Ana Protić, Zhigang Hao, Vivian Liu, Jake Salerno, Yu Wang, Mania Bankova, Long Pan, Erwin Rosenberg, Bernhard Klampfl, Robert D. Müller, Serban C. Moldoveanu, Victor David, Robert Owen Bussey III, Vu Dang Hoang

First published in London, United Kingdom, 2023 by IntechOpen
IntechOpen is the global imprint of INTECHOPEN LIMITED, registered in England and Wales, registration number: 11086078, 5 Princes Gate Court, London, SW7 2QJ, United Kingdom

British Library Cataloguing-in-Publication Data
A catalogue record for this book is available from the British Library

Additional hard and PDF copies can be obtained from orders@intechopen.com

Novel Aspects of Gas Chromatography and Chemometrics
Edited by Serban C. Moldoveanu, Victor David and Vu Dang Hoang
p. cm.
Print ISBN 978-1-80356-836-2
Online ISBN 978-1-80356-837-9
eBook (PDF) ISBN 978-1-80356-838-6

Meet the editors

Dr. Serban C. Moldoveanu graduated from the University of Bucharest, Romania, with a Ph.D. in chemistry and an MS in mathematics. He held a variety of teaching positions at the University of Bucharest before emigrating to the United States in 1983, where he taught at the University of Georgia, the University of Louisville, and Mercer University. He also has extensive industrial experience working as an analytical chemist. His research focuses on various aspects of chromatography and sample preparation. He is the author of more than 150 original papers, 12 books, a number of book chapters, and several patents. He is a member of the editorial boards of the *Journal of Analytical Methods in Chemistry* and *Frontiers in Chemistry.*

Dr. Vu Dang Hoang received his Ph.D. in pharmaceutics from the University of Strathclyde, UK, in 2005. He has been lecturing in the Faculty of Analytical Chemistry and Drug Testing at Hanoi University of Pharmacy, Vietnam, since 2007, and became an associate professor in drug quality control in 2015. His expertise is in the physicochemical characterization of topical drug delivery systems and chemometrics-based methods for the analysis of drugs in pharmaceutical dosage forms and biological fluids. He also researches the integration of the rigor of quantum chemical calculations to gain insight into molecular mechanisms. He has authored more than 50 papers and edited four books on analytical chemistry.

Professor Victor David graduated from the Faculty of Chemistry, Bucharest, Romania. Between 2007 and 2019 he was the head of the Analytical Chemistry Department at the University of Bucharest, where he is now Emeritus Professor. His main field of research is separation science (theory and applications). His list of publications includes six books for Elsevier and ten for Romanian publishing houses, 12 chapters in books and encyclopedias, and 150 papers in ISI journals. From 2017-2020 he was Associate Editor of the *Journal of Liquid Chromatography and Related Technologies.* Now he is a member of the editorial boards of *Biomedical Chromatography, Molecules, Journal of Chemistry, Journal of Essential Oil-Bearing Plants,* and *Revue Roumaine de Chimie*. He is co-editor of *Analytical Liquid Chromatography – New Perspectives* (IntechOpen, 2022).

Contents

Preface

Although gas chromatography and chemometrics are both mature fields of analytical chemistry, progress is continually being made in these important fields. New technologies, new methods, and new applications have been frequently reported in peer-reviewed literature, in manufacturer catalogs, and on the internet. A number of such novel aspects are presented in this book.

The introductory chapter describes more recent developments in gas chromatography, and also the utility of chemometrics approaches in gas chromatographic analysis. Chapter 2 describes the principle of negative thermal gradient chromatography, the advantages of this technique, and its applicability. Chapter 3 presents the main characteristics and utility of portable gas chromatography/mass spectrometry systems, and discusses some specific applications. Chapter 4 describes various sampling procedures used to make flavor and fragrance samples amenable to gas chromatographic analysis. Chapter 5 discusses various new applications of gas chromatography in the analysis of biotics and xenobiotics, such as volatile compounds of biological origin, components of biological fluids, drug metabolites, and toxicants. Chapter 6 considers the use of conference matrices as an alternative to other types of screening experiments used in chemometrics to separate key variables from those that are unimportant in large sets of influential parameters. Chapter 7 looks at quantitative structure-retention relationship (QSRR) models for liquid chromatography method development.

The goal of this book is to increase understanding of the subject by including the most recent information described in a unified form by specialists. The book is addressed to a large audience, including analytical chemists in general, either working on applications or lecturing in analytical chemistry.

Serban C. Moldoveanu
R.J. Reynolds Tobacco Co.,
Winston-Salem NC, USA

Vu Dang Hoang
Hanoi University of Pharmacy,
Hanoi, Vietnam

Victor David
University of Bucharest,
Bucharest, Romania

Chapter 1

Introductory Chapter: Novel Aspects in Gas Chromatography and Chemometrics

Vu Dang Hoang, Victor David and Serban C. Moldoveanu

1. Introduction

Gas chromatography and chemometrics are important topics of analytical chemistry. They are both mature areas of research, and for this reason, the more recent progress made in these fields is not necessarily revolutionary. Nevertheless, the progress continues. A wide range of applications associated with continuous demands to improve analytical techniques is also reflected in the progress seen in gas chromatography and chemometrics.

2. An overview of progress in gas chromatography

The first gas chromatographic separations were performed more than 75 years ago [1, 2]. These separations used hydrogen as a carrier gas, a chromatographic column containing silica gel on activated carbon, and a thermal conductivity detector [3]. Important developments followed, such as the invention of flame ionization and electron capture detectors, the introduction of temperature gradient for the GC separation, the connection of a gas chromatograph with a mass spectrometer, the introduction of open-tubular (capillary) column, the introduction of capillary columns made from fused silica, etc. From the beginning of GC and up to this day, the progress in the nature of the stationary phase was also made. The first stationary phases in the packed column were made of solid porous support coated with a high-boiling fluid or porous plastic, and those were followed by the capillary columns with a bonded, cross-linked coating [4]. The minicomputer revolution allowed the introduction of computer control of the gas chromatographic instrumentation and the data processing in GC. Throughout its history, numerous other important improvements were made in gas chromatography. Among these can be mentioned the invention of other types of detectors, development of various injection procedures allowing large volume injections or cold on-column injection, development of solid-phase micro-extraction (SPME), introduction of autosamplers, development of comprehensive two-dimensional GC, introduction of fast gas-chromatography, etc.

Modern gas chromatography is strongly associated with the use of gas chromatography-mass spectrometry, which has a mass spectrometer as a detector for a GC system. As a result, the progress in mass spectrometry has been very important for the utilization of GC, and GC-MS became the most utilized and powerful technique for compound identification in mixtures, and for the detection and quantitation of trace components when they are volatile. For this reason, the progress in GC is strongly associated with the developments in MS. Besides, the important improvements in mass spectrometric sensitivity, stability, and mass range, other examples of the progress include the development of large mass spectral libraries, introduction of GC-MS/MS systems, hyphenation of GC with high-resolution MS instruments, etc.

Over the years of utilization, significant effort has been made to adjust by chemical derivatization of the analytes in order to make them volatile and stable at a higher temperature such that to extend the range of compounds capable to be analyzed by GC and GC/MS [5]. Another extension of GC and in particular of GC-MS utilization is that on polymeric materials analysis. Polymeric compounds cannot be analyzed directly by GC due to their lack of volatility, but can be analyzed after thermal decomposition using a pyrolyzer.

The previously indicated developments are far from covering many developments in gas chromatography, mainly related to incremental developments. These developments are not seen as "milestones" although they can be very useful for practical purposes. For example, in the development of chromatographic columns for GC, among such incremental improvements is the decrease in column bleeding. As the temperature of the GC oven increases, some compounds from the stationary phase start to be generated and in the mass spectrometric detection produce an undesirable background. An important progress in making stationary phases aimed for the reduction of column bleeding allowing a better detection and also extending toward higher values the range of temperatures at which the columns can be utilized. Instrumentation in gas chromatography also has experienced numerous incremental improvements. Among these can be indicated the better control of oven temperature, better pneumatic control of pressures in the instrument, replacement of gas cylinders with gas generators, introduction of gas switches for helium conservation, development of portable GC (and GC-MS) systems, etc.

Another aspect related to the progress in gas chromatography refers to the extension of a range of applications for this technique. From the initial applications of separating gases and highly volatile compounds, gas chromatography continually expanded its range of applications. This included numerous applications for oil industry as well as for the analysis of fragrances and flavors, environmental pollutants (air, water, soil, etc.), pharmaceutical drugs, compounds in food, beverages and agricultural products, etc. A special role of GC (and GC-MS) is in the analysis of biological samples such as breath condensate, volatiles emitted from skin or bodily fluids, various xenobiotics, etc. A large part of biological samples is, however, non-volatile or highly polar molecules that cannot be directly analyzed by GC. Because of the importance of biological sample analysis, a significant effort has been made to process this type of sample by transforming them into volatile/semi-volatile compounds amenable for GC analysis.

Comprehensive two-dimensional gas chromatography coupled to mass spectrometry (GC × GC-MS) is now a common analytical technique used for the study of various complex samples, and the chemometrics based approaches are designated to decode the large amount of analytical information produced by this process (e.g. [6]).

A very large body of publications including papers in peer-reviewed journals, books, manufacturer catalogs, and information on the internet cover novel aspects in gas chromatography. Several novel items from a long list of improvements in the field are discussed in the present book.

3. Trends in chemometrics-based GC analysis

Historically, the application of chemometrics to GC analysis probably commenced in the 1990s with the works of Mayfield et al. employing a chemometrics software package to classify orange essence oil varieties analyzed by GLC (with FID and MS detectors) [7] and Jurášek et al. on the use of a chemometric detector (i.e., a computer method to selectively detect isotope cluster patterns in a time series of mass spectra of GC-MS analyses) [8].

Over the past decades, the proliferation of perceptibly sophisticated analytical systems coupled with more powerful detection techniques applied to the separation of complex samples has indeed requested highly efficient data analysis and optimization strategies. Chemometrics methods play a vital role in revealing the chemical information/knowledge hidden in high-dimensional datasets acquired from multidimensional separations. The idea of dimensionality of a mixture of compounds was defined by Giddings as "the number of independent variables that must be specified to identify the components of the sample" [9]. To improve the resolution and separation power of an analytical method, Giddings also suggested that a sample is subjected to a number of different separation mechanisms (i.e., the dimensionality of a separation method) [10]. With reference to gas chromatography, heart-cutting (i.e., conventional) techniques were initially proposed as multidimensional separations, in which the effluent from the first column was fractionally injected (i.e., a fraction or several consecutive fractions) onto a secondary column coated with a different type of stationary phase [11]. Although these conventional techniques could be useful in some cases when allowing additional separation of a modest number of the target analytes in specific regions of a single GC chromatogram, such analysis is not popular for complex matrices nowadays. The dimensionality of GC analysis has thus been mainly demonstrated by the application of GC-MS and comprehensive two-dimensional gas chromatography (GC × GC). For the former, although GC-MS is commonly employed in the analytical sciences [12] many readers may not realize that it is, in effect, a multidimensional technique as a mass spectrometer as the detector adds a dimension to that of the chromatographic separation. For the latter, GC × GC is, in fact, a separation in which many sequential heart-cuts are further separated in a second column i.e., the effluent existed from the first column is periodically sampled in such a manner to preserve the separation in the first dimension and to subject all the compounds in a sample to both separation dimensions. To do so, the system must contain two orthogonal GC columns integrated by a special interface (modulator) [13].

It is worth mentioning that multidimensional GC data can have the second-order advantage for quantification using calibration standards containing only the analytes of interest without prior knowledge of possible matrix interference [14]. This can also considerably shorten analysis time by eliminating the need to resolve chromatographic overlapping signals. In practice, however, the full advantage of a combination of complementary techniques has not been efficiently exploited. For instance, GC-MS data are usually analyzed using the mass spectrometer as a filter to generate an entire chromatogram for single ions or using the GC chromatograph as a filter to identify particular peaks based on mass spectra matching. The fact that either filter is completely selective, necessitates the requirement for chemometric tools to extract all valuable information from huge amounts of multidimensional data sometimes referred to as "a tsunami of data" or more generally "Big data".

Basically, the application of chemometrics to GC analysis can be categorized into two main groups: data preprocessing and data analysis. The aim of data processing

Chemometric tools	Overall features	Ref.
	Data Preprocessing	
Baseline correction		
Savitzky-Golay	Using a polynomial fit of mth order to $(2n + 1)$ neighboring points (inclusive of the point to be smoothed) with $n \geq m$	[15]
Asymmetric least squares	Using a smoother with deviations asymmetrically weighted to estimate a baseline	[16]
Polynomial fitting	Using a polynomial of nth order that is a best fit (in a least-squares sense) for the data	[17]
Penalized least squares	Balancing the fit of a model to the data generated by the sum of squares against its roughness by altering a smoothing parameter, provided that the location of peaks in a chromatogram is established	[18]
Moving window minimum value	Sliding a window of length k across neighboring points (inclusive of the point to be smoothed) to give an array of local k-point centered minimum values.	[18]
Local minimum values coupled with robust statistical analysis	Using a linear interpolation to estimate the local minimum values in a chromatogram as a new baseline vector, with the help of a robust statistical strategy to detect outlier data points (corresponding to the unseparated peaks)	[18]
Retention-time-alignment		
Correlation-optimized warping	Dividing chromatograms into several local regions to be iteratively stretched and compressed until the Pearson correlation coefficient between the test chromatogram and the reference chromatogram is maximized	[19]
Local minimum value	Using a linear interpolation to predict the baseline after finding local minimum values in a chromatogram and eliminating outlier data points by an iterative optimization	[20]
Automatic peak detection and background drift correction	Accepting a signal as a true peak if (i) the absolute value of its first-order derivative is five times larger than a noise threshold and (ii) its second-order derivative crosses the zero-line fewer than eight times; correcting background drift by replacing regions containing peaks by linear baselines and using three-point moving-window averaging for denoising	[21]
	Data Analysis	
Classification		
	Unsupervised Pattern Recognition	
Cluster analysis (CA)	Grouping objects into clusters according to their similarity (proximity); monitoring the correctness of clusters and detecting deviation points	[22]
Hierarchical cluster analysis (HCA)	Creating a classification hierarchy that starts with each object in a single cluster and puts together clusters until only one is left	[23]
Principal component analysis (PCA)	Reducing the dimensionality of an original data set and creating new dimensions of data by conversion of strongly correlated input variables into uncorrelated values, called principal components; the first components can represent the maximum variance direction in the data and the omission of the remaining components does not result in a significant loss of information.	[23]
	Supervised Pattern Recognition	
Linear discriminant analysis (LDA)	Data are projected from a D dimensional feature space down to a D′ (D > D′) dimensional space to maximize inter-class variability and reduce intra-class variability	[24]

Chemometric tools	Overall features	Ref.
Quadratic discriminant analysis (QDA)	An LDA closely related algorithm that omits the assumption of equal covariance for all classes, but maintains the assumption of normality (unsuitable for very small sample sizes)	[25]
K-nearest neighbor (KNN)	A non-parametric algorithm that uses proximity to make classifications about grouping an individual data point. The classification of an object is based on a plurality vote of its neighbors when assigning it to the class most common among its *k* nearest neighbors (*k* is a positive integer, selected by cross-validation procedures to ensure the lowest classification error)	[26]
Random Forest (RF)	An ensemble learning algorithm that constructs multiple decision trees to find the best split to subset the data	[26]
Soft independent modeling of class analogy (SIMCA)	Samples are analyzed by PCA, with only the significant components retained. They can be identified as belonging to multiple classes, not necessarily classified into non-overlapping classes.	[27]
Support vector machine (SVM)	A linear classifier based on the kernel function can be used for non-linearly separable data by implicitly mapping them into higher dimensional feature spaces.	[28]
Partial least squares discriminant analysis (PLS-DA)	A discriminatory variant of Partial Least Squares regression	[29]
Quantification		
Partial least squares regression (PLSR)	A multivariate calibration method that reduces the predictors to a smaller set of uncorrelated components for least squares regression performance. It is very useful when the predictors are highly collinear and can be measured with error.	[30]
Artificial Neural Networks (ANN)	It is a subset of machine learning, constructed by using a set of algorithms that mimic the behavior of the human brain for pattern recognition. ANN are comprised of node layers; each node (aka. Artificial neuron) connects to another to form an extensive network for exchanging messages. It can be used as a prognostic model.	[31]
Multivariate curve resolution-alternating least squares (MCR-ALS)	Multivariate curve resolution, similar to PCA, seeks solutions accounting for the most variation possible with non-negativity constraints. Using a constrained Alternating Least Squares algorithm, MCR-ALS solves the MCR basic bilinear model.	[32]
Parallel factor analysis (PARAFAC)	It decomposes multidimensional arrays into component matrices and commonly uses ALS to calculate the decomposition.	[33]
Generalized rank annihilation (GRAM)	It is for solving an eigenvalue problem by using two data matrices simultaneously (unknown and calibration). The introduction of factor analysis is to project a target bilinear matrix onto another PC bilinear matrix space.	[34]

Table 1.
Chemometric tools for preprocessing and analyzing GC data.

strategies is to render GC data ready for accurate identification and quantification. They include (i) background correction (i.e., denoising and smoothing for the removal of low-amplitude signals irrespective of their frequency and high-frequency signals irrespective of their amplitude; drift correction for subtracting the baseline shape from a measurement) and (ii) retention time alignment (i.e., correcting inter-run variation in retention time for similar samples). After preprocessing data, the translation of complex data for a sample into useful information covers a series of steps such as (i) peak detection (i.e., locating true signals in a chromatogram), (ii) information extraction (i.e., applying data dimension reduction), (iii) classification (i.e., discriminating between sample classes with different chemical characteristics). **Table 1** displays commonly used chemometric tools fo and analyzing GC data with some typically applied studies for illustration purpose.

4. Conclusions and future outlook

In the literature, GC has been undoubtedly proven to be one of the most sensitive and popularly applied techniques for the separation and analysis of volatile and semi-volatile organic compounds. Its application can be adopted in a wide range of analytical studies in the biomedical, pharmaceutical, forensic, environmental and food sciences. To enhance the power for maximum sensitivity and selectivity, the use of multidimensional GC (especially hyphenated with mass spectrometry) is a helpful hint for sure. The analysis that relies on such modern GC techniques can generate a very powerful data platform, e.g., the use of profile data leads to more accurate information than centroid spectra. Thus, it necessitates robust data analysis strategies to extract relevant information from such GC data. Although there have been many interesting developments in the field of chemometrics-based GC analysis, it is still difficult to judge which algorithms can give the best results in general. This is because most chemometric methods were reported when addressing a specific challenge in a data set and comparisons with other approaches were infrequently sighted. It is suggested that more comprehensive studies should be done with different types of data and algorithms to shed light on the pros and cons of each chemometric tool.

In the 21st century, Artificial Intelligence (AI) is a fast-augmenting sector that has dramatically changed many aspects of daily life worldwide. It is unsurprising that AI, if exploited correctly as demonstrated e.g., [35–37], can help scientists achieve unimaginable breakthroughs and solutions regarding GC data analysis in the future.

DOI: http://dx.doi.org/10.5772/intechopen.109943

Author details

Vu Dang Hoang[1]*, Victor David[2] and Serban C. Moldoveanu[3]*

1 Faculty of Analytical Chemistry and Drug Testing, Hanoi University of Pharmacy, Hanoi, Vietnam

2 University of Bucharest, Romania

3 R.J. Reynolds Tobacco Co., United States of America

*Address all correspondence to: hoangvd@hup.edu.vn; moldovs@rjrt.com

References

[1] Prior F. Determination of Adsorption Heats of Gases and Vapors by Application of the Chromatographic Method in the Gas Phase, Doctoral thesis. Austria: University of Innsbruck; 1947 (in German)

[2] Ettre L. The beginnings of gas adsorption chromatography 60 years ago. LC-GC North America. 2008;**26**(1):48-60

[3] Bobleter O. Exhibition of the first gas chromatographic work of Erika Cremer and Fritz Prior. Chromatographia. 1996;**43**(7):444-446

[4] Jennings WG, Poole CF. Milestones in the development of gas chromatography. In: Gas Chromatography. Second ed. Amsterdam: Elsevier; 2021. pp. 1-17

[5] Moldoveanu SC, David V. Derivatization methods in GC and GC/MS. In: Kusch P, editor. Gas Chromatography: Derivatization, Sample Preparation, Application. London, UK, London: IntechOpen; 2018. DOI: 10.5772/intechopen.81954

[6] Stefanuto P-H, Smolinska A, Focant J-F. Advanced chemometric and data handling tools for GC×GC-TOF-MS: Application of chemometrics and related advanced data handling in chemical separations. TrAC Trends in Analytical Chemistry. 2021;**139**:116251

[7] Mayfield HT, Bertsch W, Mar T, Staroscik JA. Application of chemometrics to the classification of orange essence oil varieties by GLC. Journal of High Resolution Chromatography. 1986;**9**(2):78-83

[8] Jurášek P, Slimák M, Košík M. Determination of isotope cluster patterns in mass spectra of GC-MS analyses by a chemometric detector. Microchimica Acta. 1993;**110**(4):133-142

[9] Giddings JC. Sample dimensionality: A predictor of order-disorder in component peak distribution in multidimensional separation. Journal of Chromatography A. 1995;**703**(1):3-15

[10] Giddings JC. Two-dimensional separations: Concept and promise. Analytical Chemistry. 1984;**56**(12):1258A-1270A

[11] Marriott PJ, Chin S-T, Maikhunthod B, Schmarr H-G, Bieri S. Multidimensional gas chromatography. TrAC Trends in Analytical Chemistry. 2012;**34**:1-21

[12] Matheson A, Botcherby L. Trends and developments in GC and GC-MS: A panel discussion on the latest advances and future developments in gas chromatography mass spectrometry (GC-MS). The Column. 2020;**16**(10):27-32

[13] Liu Z, Phillips JB. Comprehensive two-dimensional gas chromatography using an on-column thermal modulator Interface. Journal of Chromatographic Science. 1991;**29**(6):227-231

[14] Booksh KS, Kowalski BR. Theory of analytical chemistry. Analytical Chemistry. 1994;**66**(15):782A-791A

[15] Mikaliunaite L, Sudol PE, Cain CN, Synovec RE. Baseline correction method for dynamic pressure gradient modulated comprehensive two-dimensional gas chromatography with flame ionization detection. Journal of Chromatography A. 2021;**1652**:462358

[16] Samanipour S, Dimitriou-Christidis P, Gros J, Grange A,

Samuel AJ. Analyte quantification with comprehensive two-dimensional gas chromatography: Assessment of methods for baseline correction, peak delineation, and matrix effect elimination for real samples. Journal of Chromatography A. 2015;**1375**:123-139

[17] Mecozzi M. A polynomial curve fitting method for baseline drift correction in the chromatographic analysis of hydrocarbons in environmental samples. APCBEE Procedia. 2014;**10**:2-6

[18] Fu H-Y, Li H-D, Yu Y-J, Wang B, Lu P, Cui H P, et al. Simple automatic strategy for background drift correction in chromatographic data analysis. Journal of Chromatography A. 2016;**1449**:89-99

[19] Zushi Y, Gros J, Tao Q, Reichenbach SE, Hashimoto S, Arey JS. Pixel-by-pixel correction of retention time shifts in chromatograms from comprehensive two-dimensional gas chromatography coupled to high resolution time-of-flight mass spectrometry. Journal of Chromatography A. 2017;**1508**:121-129

[20] Fu H-Y, Hu O, Zhang Y-M, Zhang L, Song J-J, Lu P, et al. Mass-spectra-based peak alignment for automatic nontargeted metabolic profiling analysis for biomarker screening in plant samples. Journal of Chromatography A. 2017;**1513**:201-209

[21] Yu Y-J, Fu H-Y, Zhang L, Wang X-Y, Sun P-J, Zhang X-B, et al. A chemometric-assisted method based on gas chromatography–mass spectrometry for metabolic profiling analysis. Journal of Chromatography A. 2015;**1399**:65-73

[22] Passarella S, Guerriero E, Quici L, Ianiri G, Cerasa M, Notardonato I, et al. Dataset of PAHs determined in home-made honey samples collected in Central Italy by means of DLLME-GC-MS and cluster analysis for studying the source apportionment. Data in Brief. 2022;**42**:108136

[23] Gilbert N, Mewis RE, Sutcliffe OB. Classification of fentanyl analogues through principal component analysis (PCA) and hierarchical clustering of GC–MS data. Forensic Chemistry. 2020;**21**:100287

[24] Zhou X, Li X, Zhao B, Chen X, Zhang Q. Discriminant analysis of vegetable oils by thermogravimetric-gas chromatography/mass spectrometry combined with data fusion and chemometrics without sample pretreatment. LWT. 2022;**161**:113403

[25] Aghili NS, Rasekh M, Karami H, Azizi V, Gancarz M. Detection of fraud in sesame oil with the help of artificial intelligence combined with chemometrics methods and chemical compounds characterization by gas chromatography–mass spectrometry. LWT. 2022;**167**:113863

[26] Yun J, Cui C, Zhang S, Zhu J, Peng C, Cai H, et al. Use of headspace GC/MS combined with chemometric analysis to identify the geographic origins of black tea. Food Chemistry. 2021;**360**:130033

[27] Becerra V, Odermatt J, Nopens M. Identification and classification of glucose-based polysaccharides by applying Py-GC/MS and SIMCA. Journal of Analytical and Applied Pyrolysis. 2013;**103**:42-51

[28] Gerhardt N, Schwolow S, Rohn S, Pérez-Cacho PR, Galán-Soldevilla H, Arce L, et al. Quality assessment of olive oils based on temperature-ramped HS-GC-IMS and sensory evaluation: Comparison of different processing approaches by LDA, kNN, and SVM. Food Chemistry. 2019;**278**:720-728

[29] Toraman HE, Abrahamsson V, Vanholme R, Van Acker R, Ronsse F, Pilate G, et al. Application of Py-GC/MS coupled with PARAFAC2 and PLS-DA to study fast pyrolysis of genetically engineered poplars. Journal of Analytical and Applied Pyrolysis. 2018;**129**:101-111

[30] Aishima T. Comparing predictability of GC-MS and e-nose for aroma attributes in soy sauce using PLS regression analysis. In: WLP B, Petersen MA, editors. Developments in Food Science. Vol. 43. Amsterdam: Elsevier; 2006. pp. 525-528

[31] Vyviurska O, Koljančić N, Gomes AA, Špánik I. Optimization of enantiomer separation in flow-modulated comprehensive two-dimensional gas chromatography by response surface methodology coupled to artificial neural networks: Wine analysis case study. Journal of Chromatography A. 2022;**1675**:463189

[32] Izadmanesh Y, Garreta-Lara E, Ghasemi JB, Lacorte S, Matamoros V, Tauler R. Chemometric analysis of comprehensive two dimensional gas chromatography–mass spectrometry metabolomics data. Journal of Chromatography A. 2017;**1488**:113-125

[33] Valverde-Som L, Reguera C, Herrero A, Sarabia LA, Ortiz MC. Determination of polymer additive residues that migrate from coffee capsules by means of stir bar sorptive extraction-gas chromatography-mass spectrometry and PARAFAC decomposition. Food Packaging and Shelf Life. 2021;**28**:100664

[34] Prazen BJ, Bruckner CA, Synovec RE, Kowalski BR. Second-order chemometric standardization for high-speed hyphenated gas chromatography: Analysis of GC/MS and comprehensive GC×GC data. Journal of Microcolumn Separations. 1999;**11**(2):97-107

[35] Baccolo G, Quintanilla-Casas B, Vichi S, Augustijn D, Bro R. From untargeted chemical profiling to peak tables – A fully automated AI driven approach to untargeted GC-MS. TrAC Trends in Analytical Chemistry. 2021;**145**:116451

[36] Bi K, Zhang D, Qiu T, Huang Y. GC-MS fingerprints profiling using machine learning models for food flavor prediction. Processes. 2020;**8**:23. DOI: 10.3390/pr8010023

[37] Matyushin DD, Sholokhova AY, Buryak AK. Deep learning driven GC-MS library search and its application for metabolomics. Analytical Chemistry. 2020;**92**(17):11818-11825

Chapter 2

Perspective Chapter: Negative Thermal Gradient Gas Chromatography

Erwin Rosenberg, Bernhard Klampfl and Robert D. Müller

Abstract

Gas chromatography is typically operated in isothermal mode for optimum separation of a mixture of compounds with a narrow boiling point range, or in temperature-programmed mode, which strives to achieve a compromise between separation efficiency and time. Temperature gradients also keep the peak widths nearly constant over a wide range of retention times, enhancing the detectability of the later eluting peaks. In this chapter, the use of negative thermal gradients for gas chromatography (NTGGC) – for the sake of simplicity, subsequently only denoted as thermal gradient-gas chromatography, TGGC – shall be discussed. (N)TGGC is achieved by producing a stationary temperature gradient along the relatively short GC column in a proprietary experimental setup that allows cooling on one end of the column and heating on the other. The sample is injected into the hot end of the GC column, and analytes move towards the colder end of the column. Along their passage through the column, they are focused by the increasingly lower temperature of the stationary phase. This leads to a focusing of the peaks as they reach the cold column end. With appropriate temperature programming, very fast (sub-minute) chromatography with excellent resolution can be achieved on short GC columns. The present contribution will both discuss the theory behind this unusual, but highly performant mode of gas chromatographic separation, and also the hardware aspects of this technique. Relevant examples will be presented which highlight both the speed and the separation power by which (N)TGGC excels in comparison with regular temperature-programmed GC.

Keywords: temperature-programmed GC, fast GC, peak focusing, temperature gradient, column efficiency

1. Introduction

Gas chromatography is without doubt the most powerful separation technique for the analysis of volatile and semi-volatile organic (and inorganic) compounds and permanent gases [1, 2]. Under optimized conditions, peak capacities of several hundred and theoretical plate numbers in the order of several ten thousands [3] can be reached with commercial set-ups, however, at the cost of extended separation times.

Figure 1.
The 'magic triangle' of chromatographic separation: It is impossible to optimize all three factors separation speed, separation efficiency and sample capacity at the same time (redrawn after [4]).

As with other forms of chromatography, also gas chromatographic operation is governed by the "magic triangle" of chromatography [4], namely the fact that it is virtually impossible to optimize speed, separation efficiency and sample capacity of a chromatographic system at the same time (**Figure 1**). This is, because the separation efficiency (expressed as number of theoretical plates, N[1]) for a capillary column is directly proportional to the column length L, and (as can be deduced from the C-term of the van Deemter Eq. (2) inversely proportional to the column diameter d_c and the stationary phase film thickness, d_f. This means that an improvement in separation efficiency is either related to an increase in separation time, or a reduction of sample capacity under normal operating conditions. Similar mutual dependencies can be derived from the interrelation of the other parameters in the van Deemter equation.

Chromatographers have therefore searched for possibilities to overcome these inherent limitations, which has led them to develop various innovative and unconventional approaches to speed up chromatography [5–8]. Among these are:

- Micro- and narrow-bore gas chromatography,
- Vacuum-outlet (low-pressure) GC,
- Direct resistive heating GC and
- Temperature gradient gas chromatography (TGGC).

While the theoretical foundations of the first three types of fast GC shall be discussed here only briefly, the discussion of the various aspects of TGGC shall be the main focus of this chapter.

1.1 Micro- and narrow-bore gas chromatography

Gas chromatography with commercial instrumentation is often performed with columns of 0.25 mm or larger inner diameter, 0.25 μm film thickness and 30 m length. These column dimensions, providing a phase ratio of $\beta = 250$ are in many cases a good starting point for further optimization [9]. The typical number of theoretical plates

[1] See list of symbols, acronyms and abbreviations at the end of this chapter.

achievable in this setup is about 3,000/m or ca. 90,000 for a 30 m column. If scaling laws are followed, a very similar resolution and number of theoretical plates can be achieved on a 10 m column with 0.1 mm ID and 0.1 μm stationary phase film thickness. If the same linear velocity of the carrier gas is maintained, an improvement by a factor of 3 in separation time is achieved. The price to pay is that the sample capacity is lower by a factor of approximately 3^3, since the volume of stationary phase is reduced by approximately this factor (a factor 3 in column length, a factor of 2.5 in stationary phase thickness and a factor of 2.5 in column inner diameter). Even with highly sensitive detectors, this factor quickly becomes limiting, and the gain in separation speed or sample throughput is offset by the loss in sensitivity and, in particular, dynamic range.

1.2 Vacuum-outlet- (low-pressure) gas chromatography

Vacuum-outlet- or low-pressure GC operation denotes an operational mode in which the column outlet is kept at sub-ambient pressure [5, 10]. While this in fact is the case for all GC/MS instruments, there is still an important difference in the operation of columns under 'normal' conditions with a vacuum detector, and the low-pressure (LP) GC operation [9]: In the former case, column dimensions are chosen such that the column inlet can be kept at positive pressure even while the column end is at vacuum (**Figure 2**). This explains why the above-mentioned column dimension (30 m × 0.25 mm × 0.25 μm film thickness) enjoys great popularity for GC/MS operation as resulting flows are in a range that is well compatible with the pumping capacity of modern quadrupole MS systems (1–5 ml min^{-1}). If shorter columns or columns of larger ID are chosen, then flow rates in excess of 5 ml min^{-1} would result, even with the inlet being kept at ambient pressure. Alternatively, a flow restriction can be placed at the head of the column which limits the column flow and causes vacuum to extend from the detector end throughout the largest part of the column in contrast to normal operation where vacuum extends only into the final fraction of the GC column [11]. Since the diffusion coefficient in the mobile phase D_m is strongly pressure-dependent (and increases inversely proportional to total pressure, Eq. (1)), the values of the Diffusion coefficient at outlet and inlet conditions ($D_{m,o}$ and $D_{m,i}$, respectively) can be related to the pressure at inlet (p_i) and outlet condition (p_o):

$$D_{m,o} \cdot p_o = D_{m,i} \cdot p_i \tag{1}$$

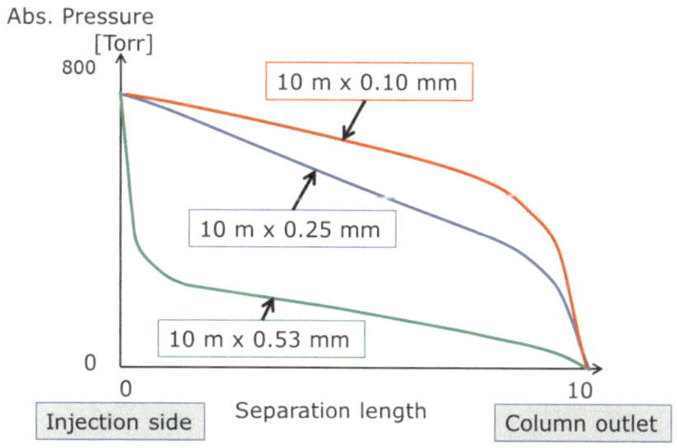

Figure 2.
Pressure drop along 10 m GC columns of different diameters. (Drawn after [11]).

It also has a pronounced effect on the terms in the van Deemter equation that depend on the mobile phase diffusion coefficient $D_{i,m}$. This relates both to the B-term (describing longitudinal diffusion) where the increasing mobile phase diffusion coefficient increases its relative contribution, as well as to the C_m-term where a larger $D_{i,m}$ reduces its contribution to the theoretical plate height (Eq. (2)):

$$H = \left[2\frac{D_{m.o}}{u_0} + \frac{11k'^2 + 6k' + 1}{96(1+k')^2} \times \frac{d_c^2 u_o}{D_{m,o}}\right] f_1 + \frac{2k'}{3(1+k')^2} \times \frac{d_f^2}{D_s} u_o f_2 \qquad (2)$$

where k' represents the capacity factor (also known as the retention factor), d_c the capillary column diameter, d_f the stationary phase film thickness, u_o the mobile phase velocity at the column outlet, $D_{m,o}$ and D_s the diffusion coefficients of the analyte in the mobile phase at the column outlet condition and in the stationary phase, respectively, and f_1 and f_2 are pressure correction factors according to Giddings [12].

Taking these two effects together, this leads to a shift of the minimum of the van Deemter curve, denoting the optimum separation velocity $\overline{u}_{opt}$ (Eq. (3)).

$$\overline{u}_{opt} = 8\frac{\overline{D}_m}{d_c}\sqrt{\frac{3(1+k')^2}{11k'^2 + 6k' + 1}} \qquad (3)$$

The optimum mobile phase velocity will thus scale with the average diffusion coefficient in the mobile phase $\overline{D}_m$, leading to an improvement of a factor of roughly 4 compared to operation at ambient pressure (**Figure 3**). As the slope of the right arm of the van Deemter curve also decreases, it is possible to obtain an even higher gain in separation speed, if one is willing to sacrifice some of the theoretically achievable separation. In contrast to the micro- and narrow-bore GC column approach, LP-GC utilizes normal- or even wide bore columns which offer a much larger maximum sample capacity Q (Eq. (4)) [5].

$$Q_s = \frac{5\pi}{2}\beta'' \frac{(1+k'_0)^2}{k'^2_0} \varrho_s \times d_f \times d_c \times H \qquad (4)$$

where β'' is a solute-liquid phase specific factor, k'_0 is the capacity factor at infinite dilution, and ρ_s is the density of the stationary phase. A further significant advantage

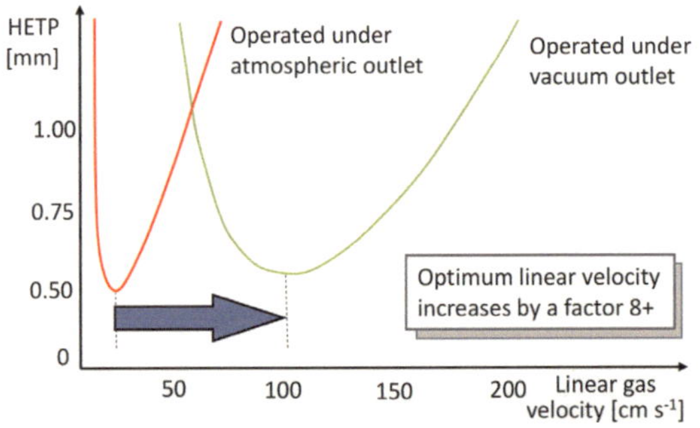

Figure 3.
Van Deemter curve for a 0.53 mm ID capillary column with He as a carrier gas for normal pressure and reduced pressure operation.

is that due to the reduced pressure within the column, the analytes elute at much lower oven temperatures (compared to normal pressure operation) which is highly beneficial for thermally labile compounds but also reduces the thermal stress to the GC column. The reduction in resolution is normally not a problem, as in most cases mass spectrometers are used as detectors that tolerate to some degree also the coelution of analytes due to their selective detection and/or signal deconvolution capabilities.

1.3 Direct resistive heating for GC

Air bath ovens are nowadays still standard in commercial instrumentation, offering operational simplicity and stability and ease of temperature control. Still, their low heating rates, high power consumption and typically bulky size do not make them the ideal choice if fast separation or portable instrumentation are envisaged. All these disadvantages can conveniently be overcome by using resistive heating which uses an electrically conductive material as the heat source [13]. To this end, the heating element either has to be placed in intimate contact with the GC column, or in the ideal case is the GC column itself. Heat is transferred by conduction or radiation. Although resistive heating was used already at a very early stage for gas chromatography [14], it was replaced soon after by air bath ovens due to their greater practicability and user-friendliness. Resistive heating only reappeared in the 1980s (although rather as a niche technique) and was continuously improved since.

Resistive heating offers fast heating and cooling rates, low power consumption and allows instruments to be built with a small footprint. All of these features make resistive heating the ideal heating technique for miniaturized and transportable GC instrumentation. Moreover, resistive heating has also become attractive for benchtop instruments where extremely fast heating rates are required that no longer can be reached by conventional air bath oven systems.

The optimal heating rate for a GC column (achieving the best compromise between separation efficiency and analysis time) is dictated by a range of parameters, such as carrier gas flow rate, column diameter and length. Blumberg *et al.* [4, 15] introduced the concepts of speed-optimized gas flow rate (SOF) and optimal heating rate ($R_{T,\ opt}$), which can be used as starting points of settings for fast GC analysis. Here, the speed-optimized flow rate is:

$$\mathrm{SOF} = f_{\mathrm{gas}} d_{\mathrm{c}} \tag{5}$$

where f_{gas}, in mL min^{-1} mm^{-1}, is determined by the carrier gas type (10 for hydrogen and 8 for helium) and d_{c} is the column internal diameter in mm. $R_{T,\ opt}$ is usually 10°C per void time [15], which results from the selected flow rate and column dimensions. Some model calculations for the optimum GC parameters and the resulting analysis times and peak capacities are reported in **Table 1**.

It becomes evident that maximizing the advantage of short column lengths for fast GC while maintaining a good peak capacity requires the operation with very high heating rates, which makes resistive column heating mandatory.

1.4 Multiplexing GC

Not a fast GC method in the strict sense, multiplexing GC still offers the possibility to increase sample throughput, and thus the number of GC runs performed in a given

Column length [m]	*SOF* [mL min^{-1}]	Void time [min]	$R_{T,opt}$ [°C min^{-1}]	Normalized peak capacity	Normalized analysis time
1	0.8	0.0134	746	31.6	4.3
3	0.8	0.0568	176	54.8	18.1
5	0.8	0.116	86	70.7	37.1
10	0.8	0.313	32	100	100

Table 1.
Optimum GC parameters and resulting separation performance for various column lengths, calculated for He as carrier gas, an internal column diameter of 0.1 mm and the void time being calculated at 50°C (after Wang et al [13]).

time frame. The idea of multiplexing GC is based on introducing a sample (either the same, or a gradually changing sample) at pre-defined intervals which are much shorter than the chromatographic run time [16, 17]. This leads to a complex chromatogram that results from the superposition of the individual chromatograms introduced at different timepoints. It is possible to deconvolute the complex chromatogram into the individual chromatograms, provided that the sequence at which the sample was introduced is known (**Figure 4**), and that it is a random, non-periodic binary sequence (0 = no sample is introduced; 1 = sample is injected).

While the concept is attractive and has in recent years been applied to gas chromatographic and other types of chromatographic [18, 19] and non-chromatographic separation [20] notably by the group of Trapp and co-workers, it requires a significant computational effort, and also can be used to monitor processes only in retrospect, as the entire data set must be recorded prior to deconvoluting the data into the original individual chromatograms.

2. Basics of gas chromatographic separation

Gas chromatography is a separation technique for compounds which are sufficiently volatile to be transported via the gas phase. As the analytes travel along the column, they encounter retention on the basis of their individually different interaction with the stationary phase and eventually are separated. The parameters that influence the resolution are the chemical nature of the stationary phase (governing the selectivity α), stationary phase thickness d_f, column length L (proportional to theoretical plate number N) and inner diameter d_c (inversely proportional to theoretical plate number), carrier gas velocity u (allowing to reach optimal, that is, minimal values for the theoretical plate height H) and the column operation temperature. Since the latter parameter is the easiest to change from the practical point of view, practical method development typically starts with the adjustment and optimization of the GC column temperature programme.

In classical gas chromatography, two major modes of operation are distinguished: isothermal GC and temperature-programmed GC. In isothermal GC, the analytes are separated at constant column temperature. This leads, for the members of a homologous series, to exponentially increasing retention times t and to peak widths W that increase roughly proportionally (proportionality factor b) with retention time (see Eq. (6)) [21].

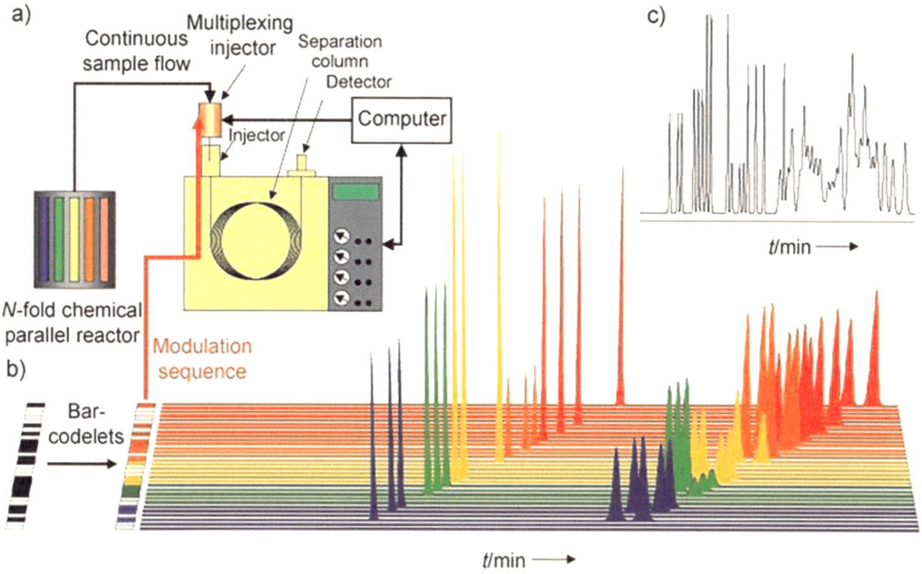

Figure 4.
Concept of multiplexing GC used for high-throughput analysis: a) schematic experimental setup for an analytical system equipped with a multiplexing injector. The samples are sequentially injected by short pressure pulses (1–5 ms) onto the separation column by the multiplexing injector according to an n-bit binary pseudo-random sequence (n = 5) with time intervals Δt on the order of seconds. b) Temporally shifted chromatograms obtained by repetitive sample injections according to the pre-determined pseudo-random sequence. c) Convoluted chromatogram, which represents the sum of the chromatograms depicted in (b). (Reprinted with permission from O. Trapp, Angew. Chem. Int. Ed. 46 (2007) 5609–5613. © 2007 WILEY-VCH Verlag GmbH & Co. KGaA, Weinheim [17]).

$$W = b.t \tag{6}$$

Although peak width increases with retention time, the resolution of two adjacent peaks improves by a factor $\sqrt{N}$, and thus with the square root of the column length, which causes a proportional increase in separation time if the measurements are performed at the same linear (average) velocity. However, the price that one has to pay is the loss in sensitivity, as the peak height becomes the smaller, the wider the peaks are. At a certain point, the signal-to-noise ratio will become so low as to prevent their detection.

While isothermal separations are always superior to temperature-programmed separations under comparable conditions in terms of achievable resolution, these are only recommended where the analytes have a relatively narrow range of boiling points to avoid unacceptably long separation times. However, as the GC system is in constant thermal equilibrium, very stable retention times and chromatographic peak areas are typically produced because the baseline, inevitably caused by column bleed, is either very low or constant throughout the entire run.

In contrast to this, a temperature gradient- or temperature-programmed separation is performed when an analyte mixture of largely different composition and boiling points is to be analyzed [22]: In that case, the change of column temperature is associated with a change in chromatographic retention, expressed by the capacity factor (often also called retention factor) *k'* (Eq. (7)):

$$k' = t_r/t_m \tag{7}$$

where the temperature dependence of k' is described by:

$$k' = k_0' \, exp \ (\Delta G/(\mathcal{R}\mathrm{T})), \text{or} \tag{8}$$

$$ln \ k' = ln \ k_0' + (\Delta G/(\mathcal{R}\mathrm{T})) \tag{9}$$

In Eqs. (8) and (9), k_0' is the retention factor of some previously chosen reference substance, $\mathcal{R}$ = 8.314 J mol^{-1} K^{-1} the universal gas constant, T the absolute temperature in K and ΔG an increment (relative to the reference solute) in Gibbs free energy of desorption of a given solute from the stationary phase.

Two analytes will be separable from each other by gas chromatography, if there exists a difference in their interaction with the stationary phase, which implies $\Delta G_1 \neq \Delta G_2$ and consequently leads to $k_1' \neq k_2'$.

At the same time, both parameters (capacity factor or retention factor k' and Gibbs free energy change ΔG) show a distinct dependence on temperature. Keeping in mind that the capacity factor k' is related to the distribution constant or partitioning coefficient K of an analyte through the phase ratio β (Eq. (10)):

$$k' = K/\beta \tag{10}$$

where the phase ratio β is defined as (Eq. (11)):

$$\beta = V_m/V_s \tag{11}$$

and V_m and V_s are the volumes of mobile and stationary phases, respectively, in the GC column, we can express the relation between the partitioning coefficient K and the change of Gibbs standard free energy $\Delta G°$ at equilibrium by (Eq. (12)):

$$\Delta G° = -\mathcal{R}T \ ln \ (K) \tag{12}$$

As it is known from thermodynamics that the change in Gibbs free energy can be related to the change of the standard enthalpy $\Delta H°$ and the standard entropy change $\Delta S°$ according to (Eq. (13))

$$\Delta G° = \Delta H° - T\Delta S° \tag{13}$$

Substituting Eq. (13) into Eq. (12) leads to:

$$\Delta H° - T\Delta S° = -\mathcal{R}T \ ln \ (K) \tag{14}$$

which can be rearranged to yield:

$$K = exp\left(-\frac{\Delta H°}{\mathcal{R}T} + \frac{\Delta S°}{\mathcal{R}}\right) \tag{15}$$

As both $\Delta H°$ and $\Delta S°$ can, in a first approximation, be considered constant for a narrow temperature interval, it becomes evident that the partitioning coefficient critically depends on temperature. Even a small temperature change can have a remarkable effect on the partitioning coefficient and hence on retention. This effect is used to maximize the difference in relative retention between analytes, and thus to effect separation.

The importance of temperature in GC separations has already been known from the beginning of its development. Already in the very first examples of successful GC separations, different column temperatures were chosen to separate different mixtures [23, 24]. However, in the early years of gas chromatography, it was experimentally difficult to reproducibly set different column temperatures and keep them constant. Thus, volatile compounds were separated at low column temperatures while less volatile compounds eluted at long retention times and with broad peaks. This problem is generally referred to as the "general elution problem" which can only be overcome by altering the capacity factor *k*' from high values at the beginning of the chromatogram to lower values towards the end of the chromatogram, achieved by an increase of column temperature. Griffiths *et al.* [25] demonstrated as early as 1952 the benefits of changing the temperature of the GC column to improve separation. However, it should take until the late 1950s for both the instrumentation and the theory for temperature-programmed GC (TPGC) as developed, largely led by the instrumental developments and the theoretical treatment of Dal Nogare [14, 26, 27]. TPGC was demonstrated to provide a solution to the general elution problem and quickly became the primary separation mode in GC.

Temperature-programmed GC (TPGC) has a number of advantages over isothermal GC (ITGC), which are [28]:

- better resolution of early eluting peaks,
- better detectability for late eluting peaks,
- shorter analysis times,
- removal of less volatile sample constituents (matrix or contaminations) from the column if the temperature is increased sufficiently at the end of the run or held for a certain period at the gradient end temperature,
- decreased peak width and hence increased peak height and enhanced sensitivity for late eluting peaks and
- better peak shapes and precision (as a result of better-defined peak boundaries).

This is contrasted by a number of drawbacks which, however, are normally by far offset by its advantages. These include:

- the need for more complex instrumentation,
- an increase in baseline noise,
- limitations to use certain stationary phases due to lack of suitability for use at high temperatures, and
- eventually, longer total analysis time due to a long cooling period after each analysis.

In TPGC, three types of temperature profiles are generally used:

- *linear profiles*, at which the temperature is changed at a constant rate,
- *multilinear profiles*, which consist of several phases of either isotherm operation or heating at a constant rate and
- *ballistic profiles*, which occur when an oven is rapidly heated. The heating rate changes over time.

3. Thermal gradient gas chromatography (TGGC)

Classical chromatographic operation modes achieve the separation of analytes either under constant retention conditions, or by initially retaining these strongly at the start of the chromatogram, and then reducing their retention by lowering the capacity factor for these compounds. In gas chromatography, this corresponds to the operational modes of isothermal GC (ITGC) and temperature-programmed GC (TPGC). In liquid chromatography, the equivalent modes would be isocratic separation (= separation under constant elution strength) and gradient separation (with a solvent of increasing elution strength). In both GC and LC separation, the application of a gradient results in the decrease of the capacity factor *k*', and hence in reduced retention. This illustrated in **Figure 5**, where the retention ratio R_r has been introduced as a dimensionless parameter describing the analyte velocity relative to the mobile phase velocity (Eq. (16)):

$$R_r = \frac{1}{1+k'} \tag{16}$$

It is a characteristic property of separation under static (isothermal/isocratic) conditions that the axial dispersion of the analyte band within the column increases with migration distance. In isothermal separation, however, the retention time difference between two (differently) retained peaks increases stronger than the peak width does. Isothermal separations represent thus the best achievable separation from a theoretical point of view. With increasing temperature, retention times become shorter (and separation consequently faster) but also resolution between two adjacent peaks is reduced. The explanation is that the migration velocity of the two analytes approaches the mobile phase velocity with increasing temperature, leading to a partial and finally a complete loss of resolution when a temperature is reached at which both analytes are exclusively in the mobile phase. For (linearly) temperature-programmed separations, however, separation benefits from the fact that the analytes are at least partially retained (and they consequently move through the column at a lower speed than the mobile phase velocity) as long as the column temperature is below their boiling point. Once the boiling point of this substance is reached, or more correctly, once it no longer partitions into the stationary phase, it starts to travel along the column with the velocity of the mobile phase. The separation of the analytes is thus achieved in the first part of the temperature gradient of the separation, where the analytes have (due to their individual affinity towards the stationary phase) different linear velocities in the column. As soon as the analytes are both only present in the mobile phase, they are transported towards the detector with a constant time offset,

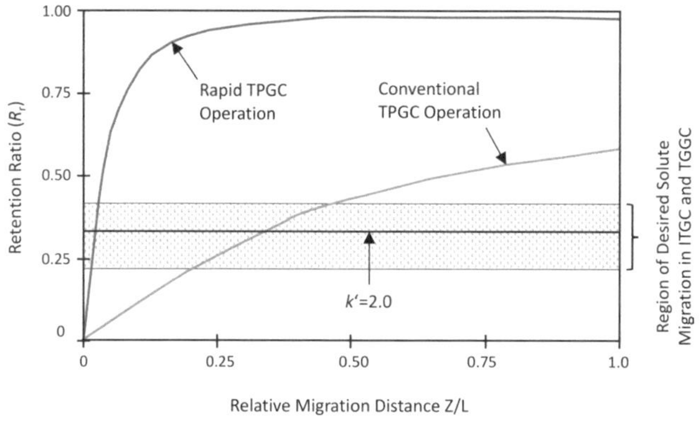

Figure 5.
Plot of retention ratio (R_r) against migration distance for different modes of chromatographic operation. Abbreviations and symbols: PTGC: Programmed-temperature GC; TGPGC: Thermal gradient-programmed-temperature GC, k': Capacity factor; L: Column length; Z: Traveled distance of the analyte (redrawn after Rubey [29]).

and resolution does, in fact, not change much. This behavior is clearly seen in the simulations reported in **Figure 6**, which represent simulated results[2] for the separation of linalool and linalool acetate under isothermal conditions at different temperatures (**Figure 6a**), and under different gradients (**Figure 6b** and **c**). In the simulation of the isothermal separation, it becomes evident that (theoretically) the best separation is achieved at low oven temperature (at the price of an excessively long duration of the separation). Increasing the isothermal temperature reduces both the absolute and the relative retention and thus decreases the resolution. For the case of temperature-programmed GC with a linear ramp, it can be seen in **Figure 6b** and **c** that for a given temperature window (separation with the same gradient steepness but different starting temperatures), the absolute difference in retention times is approximately constant, and so is also the peak width. Consequently, the resolution is constant in this window of operating temperatures, and it only starts to decrease when the gradient starting temperature is so high that the second, slower analyte is no longer retained sufficiently relative to the first, faster-traveling analyte. At a certain point, the resolution of the considered peak pair is lost or at least significantly compromised. The steeper the temperature gradient is, the earlier this point is reached (compare **Figure 6b** and **c**).

Temperature gradient GC separations are different from ITGC and TPGC in that a temperature gradient is applied; however, this gradient is normally a gradient in space rather than in time, and the gradient leads to a decrease of column temperature in axial direction. While the use of such a gradient is contra-intuitive according to normal chromatographic separation theory, it bears a number of advantages over classical modes of operation as the axial negative thermal gradient leads to a reduced *k*' value and hence a reduced migration velocity of the analyte peaks as they travel down

[2] Simulations of GC retention time and peak width were performed with the web version of the freeware programme "Restek Pro EZGC Chromatogram Simulator" (available at: https://ez.restek.com/proezgc, accessed on 10.01.2023). Simulations were performed for the analytes linalool and linalool acetate on a Rtx-1 column of dimensions 30 m × 0.25 mm × 0.25 μm with a flow of He at 2 ml min^{-1} under the conditions specified.

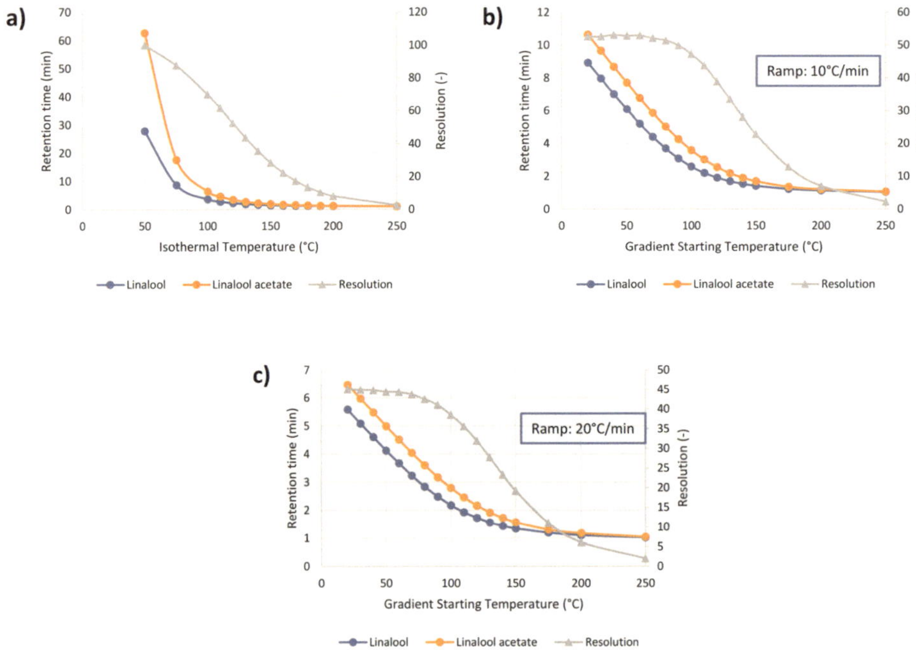

Figure 6.
Simulations of gas chromatographic retention and the resulting resolution for linalool and linalool acetate (simulated using the Restek Pro EZGC Chromatogram Simulator [https://ez.restek.com/proezgc,]) under a) isothermal conditions, b) gradient separation conditions at different starting temperatures and a ramp of 10°C/min and c) gradient separation conditions at different starting temperatures and a ramp of 20°C/min.

the column. Since the leading edge of a peak is decelerated versus its centre or trailing edge, the peak is focused (**Figure** 7).

Remarkably narrow (and thus high and well-detectable) peaks can result from this mode of operation; however, as all peaks are decelerated in a static temperature gradient GC (TGGC) system, the resolution is typically also reduced in comparison to an isothermal separation system.

In fact, it has been a matter of debate whether peak focusing can improve the resolution of a negative thermal gradient system over the best achievable isothermal separation (called by Blumberg '*idealized basic separation*', IBS) [31–33]. The conclusion was that even peak focusing through a negative thermal gradient could not improve resolution over what is achievable in the IBS under isothermal conditions [34]. However, as the resolution is often limited by practical problems (slow injection, cold spots, peak tailing), TGGC counteracts many of these and is thus capable of bringing the resolution of practical chromatograms closer to the theoretically achievable performance limit [35].

To overcome the limitation of decreasing resolution as the peaks get slower as they move towards the (colder) column end, TGGC can also be operated in the dynamic mode. This operation mode involves using an axial negative thermal gradient along the GC column, which is ramped during the chromatogram. The increase in temperature as a function of time prevents the analytes from getting stuck on the column, thereby losing the separation already achieved. Important parameters that govern the resolution are the speed at which the temperature is ramped up and the steepness of

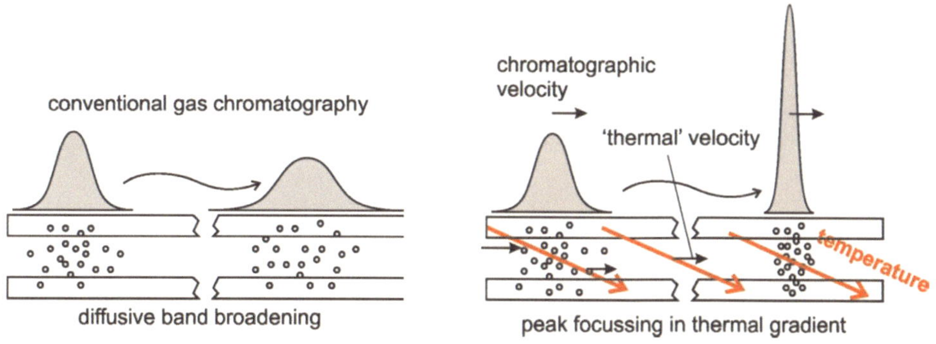

Figure 7.
Peak broadening in conventional GC (left) compared to peak compression due to the negative temperature gradient (right). In equilibrium, the thermal velocity of the sample is identical to the chromatographic velocity of the sample. (Reprinted with permission from P. Boeker, J. Leppert, Anal. Chem. *87 (2015) 9033–9041. © 2015 American Chemical Society [30]).*

the gradient (temperature difference between the inlet and outlet of the GC column). The different modes of chromatographic operation are listed in **Table 2** and graphically represented in **Figure 8**.

Thermal gradient gas chromatography can offer several advantages over PTGC. The most important ones are listed below [36]:

- Use in hyperfast-GC analyses possible,
- Elution at lower temperatures than in PTGC, especially useful for the analysis of thermally unstable substances (**Figure 9**),
- Increased chromatographic resolution,
- Possible use in miniature and micro-GC units,
- Continuous sample injection is possible in some designs.

Producing and maintaining a stable thermal gradient requires a very different and dedicated instrumental setup than conventional GC. The various approaches to designing and constructing instrumentation capable of TGGC mode operation will be discussed in the subsequent section. An overview of the different possibilities for creating a temperature gradient along the column is given in **Figure 10** (after Conteras [36]).

		$T = f$(Position along the column)?	
		No	**Yes**
$T = f$(Time)?	**No**	Isothermal GC (ITGC)	TGGC with a stationary gradient
	Yes	Programmed-temperature GC (PTGC)	TGGC with a moving gradient

Table 2.
Modes of chromatographic separation.

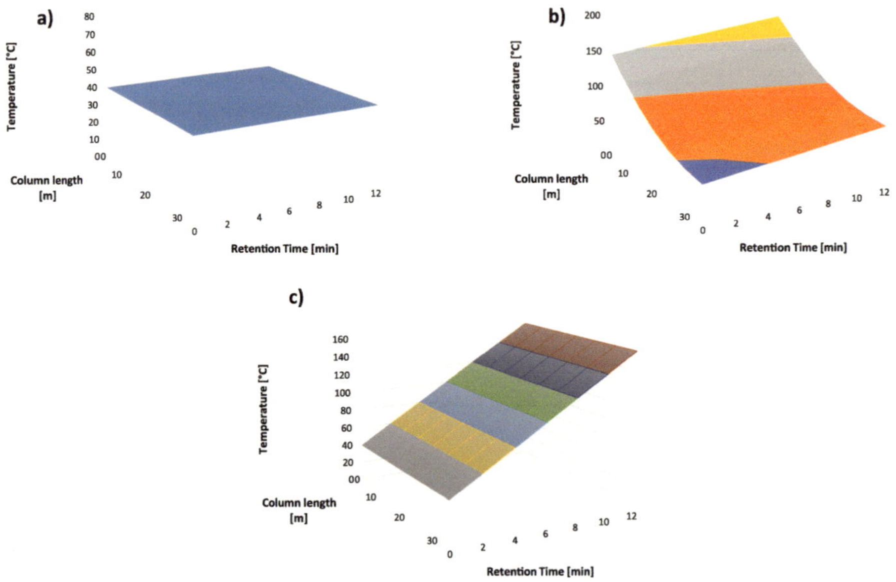

Figure 8.
Graphical representation of the different modes of chromatographic separation, characterized by their temperature profiles as a function of position along the column and retention (separation) time for a) isothermal separation (ITGC), b) thermal gradient GC (TGGC) and c) programmed-temperature GC (PTGC).

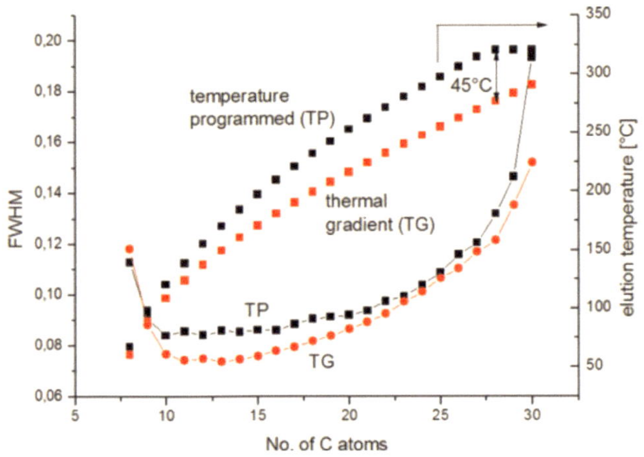

Figure 9.
Illustration of one of the most important advantages of TGGC: The elution of analytes (here: n-alkanes C8 – C30) at significantly lower column temperature than in TGGC. The difference can be as large as 45°C, as shown here, while the peak width of TGGC and PT-GC separations is practically the same. (Reprinted with permission from P. Boeker, J. Leppert, Anal. Chem. *87 (2015) 9033–9041. © 2015 American Chemical Society [30]).*

4. Producing axial temperature gradients in GC

Very soon after the establishment of gas chromatography as a versatile separation technique [37], different operational modes were studied to improve its performance.

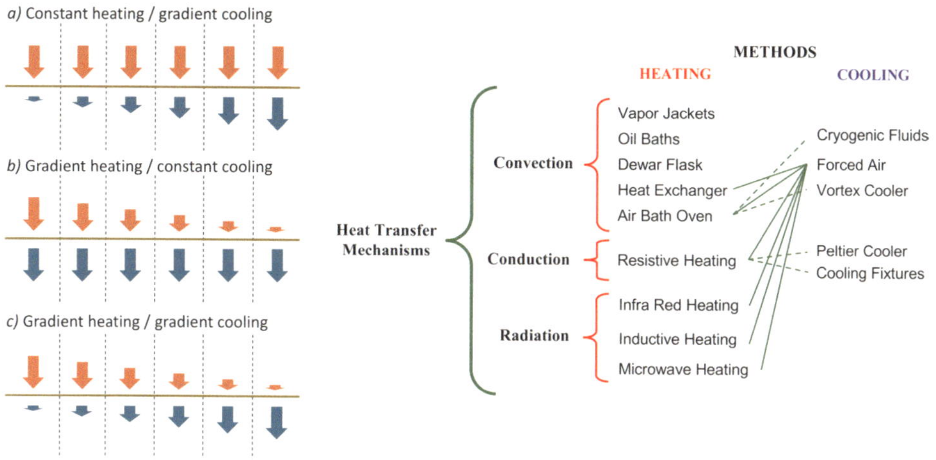

Figure 10.
Left: Possibilities of creating thermal gradients along a GC column. Right: Classification of heating and cooling methods by heat transfer mechanism (after Contreras [36]).

The first application of axial thermal gradients in gas chromatography was probably reported by Zhukovitskii *et al*. in 1951 [38]. In this work, a furnace was used to generate a temperature gradient between the head of the column that was heated to the highest temperature and where the heat dissipated towards the end of the glass column, filled with a solid adsorbent. This method was capable of reducing the severe peak tailing seen in isothermal operations. This variant of gas chromatography was named 'chromathermography', a name also used later on by several other groups [39]. In 1956, Zhukhovitskii introduced a modification of the original design which had a furnace that was moving along the packed GC column to create a dynamic temperature gradient [40]. In this design, frontal chromatography was combined with a non-stationary gradient to allow the semi-continuous analysis of samples [41]. Further developments of this principle became in the late 1950s and early 1960s a mainstay for chromatographic analysis in the USSR, with two commercial instruments, namely models KhT-2 and later KhT-2 M, being introduced on larger scale [42–44]. In these instruments, both active heating and cooling were implemented, the former being achieved by contact heating of the coiled chromatographic column and the latter by blowing cool air counter-currently to the direction of the carrier gas stream in the column. Relatively little notice was taken outside the USSR of this technique, mainly because hardly any publication was available outside the USSR [38]. Tudge reviewed this and several other Russian papers related to chromathermography and contributed to this technique's theory [45]. In the USA, Nerheim published a paper on this method [46]. This work generated a heated zone by a glass sleeve wrapped with heating tape that was moved along a short linear glass GC column. The oven was passed several times over the GC column whereby the temperature was increased from one passage to the next, allowing the separation of individual peaks. The next contribution to this type of chromatography was made by Ohline and DeFord, who used a 15″ long oven consisting of an aluminum bar that was heated on one and cooled on the other end [37]. Due to the use of cooling (cold water) and heating medium (steam), only relatively low-temperature gradients of 1 to 8.5°C/cm could be reached. In addition to a theoretical comparison of separation times in ITGC and TGGC, this allowed an acceptable separation of low alkanes (C5-C9).

In the early 1970s, a new way of creating the thermal gradient was introduced by Vergnaud and co-workers [47, 48] which in fact marked the transition from what was hitherto called 'chromathermography' (where the heated zone is moved along the GC column, and consequently the temperature gradient is created only along a (short) section of the entire GC column) to thermal gradient GC where the temperature gradient extends along the entire chromatographic column which however still is of short length (typically below 5 m, and more often even below 2 m). The temperature gradient was created by resistive heating with a heating wire coiled around the separation column [49, 50]. With this general idea, various operational modes were available, such as isothermal and programmed-temperature operation, temperature gradient operation and also backflushing, depending on the control of the heated zones [47, 48]

While this experimental setup already provided considerably increased flexibility as compared to conventional operational modes in GC, this was taken even further by the approach of Fenimore [51], who designed an experimental setup in which column sections would be heated individually. To this end, an 11 m long capillary column was coiled around five sections of brass tubing of 4.35 cm OD, which could be heated individually by coiled nichrome heating elements mounted on ceramic tubing coaxial to the brass tubing. Each coil held, within a grove machined into the surface of the brass tubing, approximately 2.25 m of column length. This allowed the separation of C10-C18 hydrocarbons in less than 3 min, and as the heaters could be controlled individually, also the use of different operational modes.

In the second half of the 1970s and early 1980s, some few papers appeared on chromathermography for preparative use and discussed the practical realization [52, 53] and the quantitative aspects of this technique, which was considered as an analogue to frontal (displacement) chromatography for liquid chromatography where the role of the displacement solvent was taken by the heater element.

After a long period of hibernation, renewed interest in the TGGC technique arose in the early 1990s. Rubey both patented [54, 55] and published [29, 56] an approach to produce axial thermal gradients where a column mounted in a sheath assembly on a heat exchanger allows establishing a temperature gradient along the column. Heating is achieved by an electrical heater that provides a constant amount of heat along the column length, while cooling is done with a stream of nitrogen that is pre-cooled when entering the heat exchanger and loses its ability to cool the column as it passes along the column in counter-current orientation to the carrier gas stream. In addition to introducing the three-dimensional view of temperature distribution along the column length and with time that we also use in **Figure 8** to illustrate the characteristics of TGGC in comparison with ITGC and TPGC, Rubey succeeded in separating a mixture of *n*-alkanes with wide volatility differences (*n*C8 – *n*C22) within 100 seconds.

In a series of papers, Jain and Phillips [57–59] developed an experimental setup for TGGC in which the temperature gradient was created by directly resistively heating the capillary GC column. This was achieved by using a thin electrically conductive film applied to the outside of the column such that a negative resistance gradient was created along the column. The negative temperature gradient along the column continuously refocused eluting bands, resulting in sharper and taller peaks. The authors also concluded that the proposed technique showed promise for rapid analyses of flowing streams and, thus, for real-time monitoring applications.

The revived interest in TGGC was also demonstrated by several patents filed in the early 1990s with different materializations of the TGGC principle. The patents of

Rubey [54, 55] were already mentioned above; they described a TGGC system in which the thermal gradient was created by controlling the temperature of a heat transfer fluid via resistive heating. In 1993, Hiller *et al.* patented a TGGC system [60] where the GC column is incorporated into a system of two coaxial tubes. Through these tubes, a cold and a hot heat transfer fluid are circulated counter-currently and allowed through a heat exchanger and the control of the fluid flow rates and temperatures the production of different temperature gradients.

Rounbuehler *et al.* filed in 1998 a patent [61] in which the production of thermal gradients by various approaches was claimed, among these using directly resistively heated metal columns of different cross-sections for the increasing removal of heat from a uniformly heated metal capillary by a more efficient heat exchanger. Although the theoretical concepts are interesting, the patent seems to be a rather hypothetical work, as the authors have not reported any chromatogram recorded with their approach, nor have they published any results obtained with any of their described systems.

Only one publication on TGGC seems to have appeared in the decade from 2000 to 2010: This is the work of Zhao *et al.* [62], who have produced a temperature gradient on a 70 cm PLOT capillary column (filled with Porapak™ Q) inserted in the spiral grove of a brass plate. This plate – and consequently the GC column seated therein – was heated by a centrally located heating element, and the temperature gradient was created by the dissipation of heat to the environment. Although only a very shallow temperature gradient (ca. 1°C cm^{-1}) could be produced this way, some improvement in separation time over classical TPGC was reached while separation efficiency remained almost unaffected.

A significant impulse to this direction of research was given by Contreras in 2004, who then submitted his Master thesis at the University of Dayton [63] that was devoted to the design and application of thermal gradient programming techniques for use in multidimensional gas chromatography–mass spectrometry (MDGC-MS). In this thesis, TGGC operation was proposed as a technique that allows focusing of the analytes eluting from the 1D-column at the head of the 2D column and their subsequent fast separation in the second dimension. To this end, a column sheath assembly was constructed to create an axial temperature gradient in the column, and have a fast heating and cooling cycle, while keeping radial temperature uniform within the column.

While rapid heating usually is not a problem in (comprehensive) MDGC, it was correctly pointed out by the thesis' author that rapid cooling is problematic, which in this case was achieved by a mechanically modulated device that exposed different sections of the second-dimension column to a liquid-nitrogen cooled stream of gaseous nitrogen. Unfortunately, none of the considerations of this author regarding TGGC operation was published outside of his Master thesis. However, in his PhD thesis, performed at Brigham Young University, Utah, under the supervision of Milton Lee, Contreras returned to the investigation of axial temperature gradients in gas chromatography, which he has already started in his Master thesis [36]. Two publications resulted directly from this PhD thesis in which Contreras discussed the possibility of using a TGGC system for fast separation.

The first of these two publications [64] describes the "peak sweeping" mode of TGGC operation. This is based on introducing a sample into a column with a preset decreasing temperature gradient along its length, waiting for a short time until the sample separates along the gradient, and then raising the temperature to sweep all of the compounds out of the column and into the detector ("peak sweeping"). To demonstrate the feasibility of this approach, a simple laboratory apparatus was

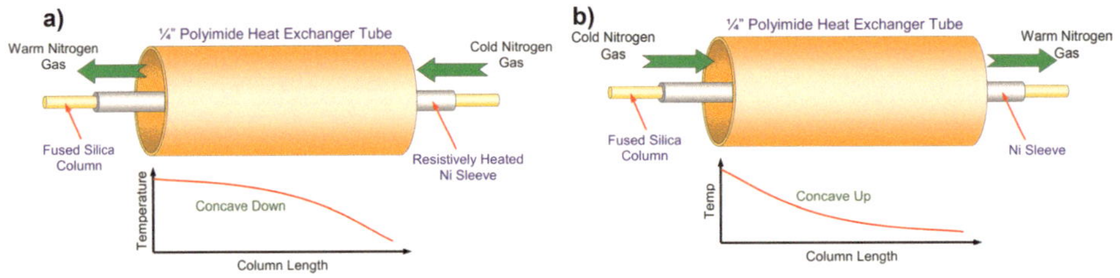

Figure 11.
Heat exchanger configuration for creating (a) concave down and (b) concave up temperature profiles along the GC column (from Contreras [36]).

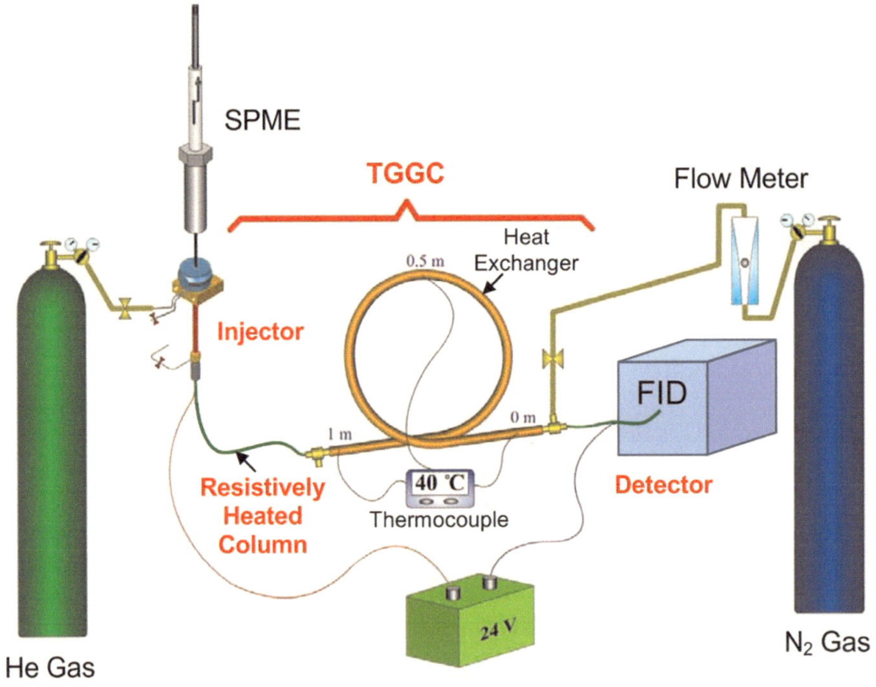

Figure 12.
TGGC system for generating axial temperature gradient profiles (from Contreras [36]).

constructed based on simultaneous resistive heating and convective cooling (**Figure 11**). Contreras could demonstrate by experimental comparison between isothermal GC (ITGC), temperature-programmed GC (TPGC) and TGGC that the result of TGGC separation is essentially equivalent to TPGC operation when using the same column length; however, narrower peaks and higher signal-to-noise-ratios are achieved in TGGC (**Figure 12**). Furthermore, TGGC helps to minimize band broadening and peak tailing that arise from non-ideal sample introduction or column adsorption. The extremely high column heating (4000°C min^{-1}) and cooling rates (3500°C min^{-1}) as an effect of the low thermal mass of the system allow for selective separation (i.e., "peak gating") of compounds in a mixture without sacrificing the resolution of earlier or later eluting compounds (**Figure 13**).

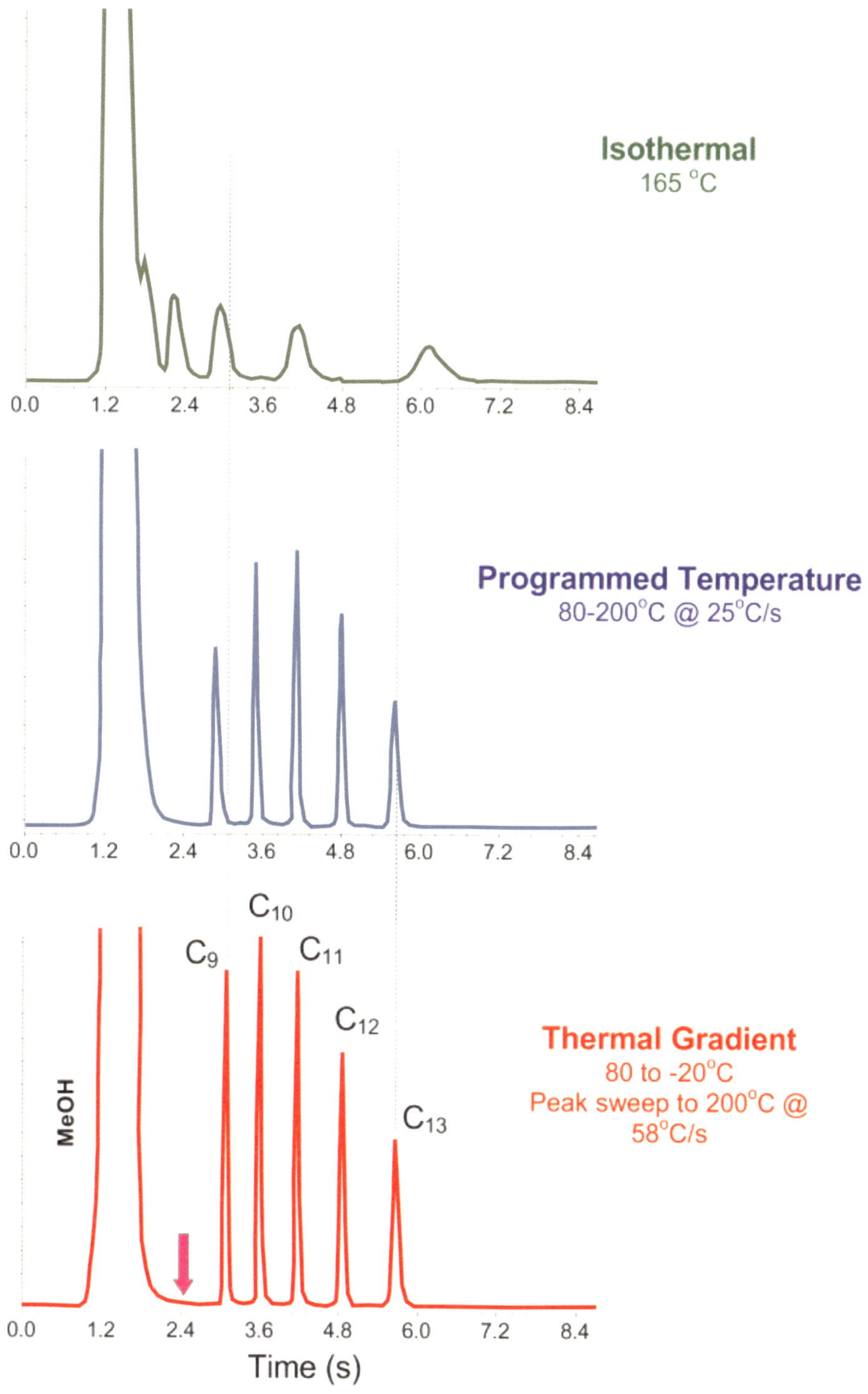

Figure 13.
GC analysis of a series of n-alkanes (C9-C13) using different separation modes. The arrow indicates when the temperature gradient was increased (sweeping) (from Contreras [36]).

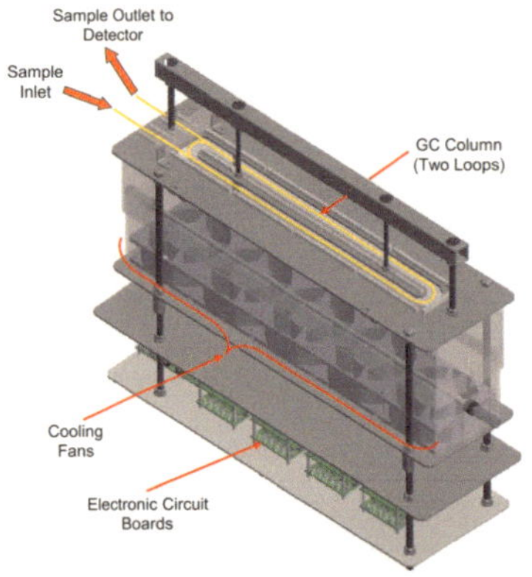

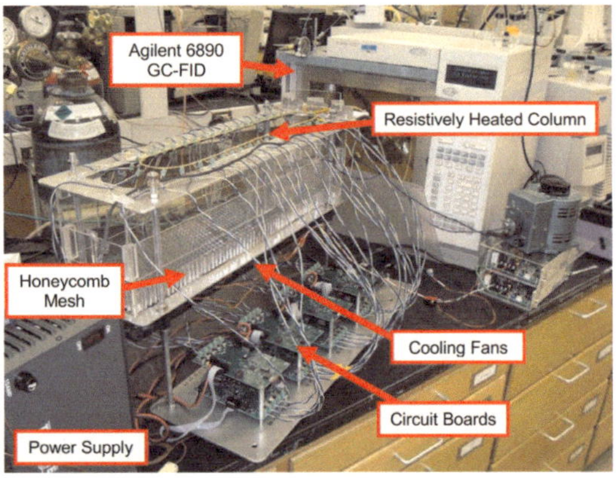

Figure 14.
Schematic three-dimensional drawing of the GC system used to create the different temperature gradient profiles (left) and photograph of the actual experimental setup (right) (from Contreras [36]).

The second paper published by Contreras in connection with his PhD thesis [65] described a TGGC system capable of rapidly producing and varying thermal gradient profiles by simultaneous use of resistive heating and convective cooling. The middle section of a 3 m GC column was inserted into a nickel tubing that was resistively heated by 40 individually addressable heated zones of each 5 cm length over an active column length of 2 m. Active cooling was achieved by five computer fans aligned along the GC column coil. The initial and terminal parts of the column were used to interface the column to the inlet and the flame ionization detector of a commercial chromatograph (**Figure 14**). Heating and cooling rates as high as 1200 and 2500°C min^{-1}, respectively, allowed the creation of dynamic temperature gradients. The separation characteristics of TGGC with dynamically changing temperature gradients were demonstrated with an experimental setup using a 1 m column length. With a gradient velocity of 2.22 cm s^{-1}, repetitive separations were possible every 45 s, and injection bandwidths of 45 s duration were transformed into peaks of approximately 1 s peak width. Dynamic TGGC enables unique control over separations, allowing to improve resolution and detection of signal-to-noise. Smart separations can be performed by TGGC in which the separation time window is most efficiently utilized, and optimized separations can be quickly achieved. However, both the energy and the space demand of this instrument are considerable, making it not very attractive in the routine laboratory, despite of its excellent chromatographic performance.

Only a limited number of further practical works related to TGGC were later on performed at Brigham Young University – among these, the PhD thesis of Wang [66] which investigated direct resistive heating and axial thermal gradients applied to microchip gas chromatography [67]. Although, due to the difficulty of microscale fabrication of the GC columns, the improvement achieved by TGGC over TPGC with regards to peak width and separation efficiency was not as impressive as at normal scale, the improvement in peak shape and the significant reduction of peak tailing was noteworthy. Instead, the group around Tolley and Lee concentrated on more theoretical studies on the separation behavior and simulation of GC separation under the different experimental conditions. These findings were published in a series of papers [68–71], many of which were based on the Master thesis of Avila published in 2021 [72]. They discuss in detail the simulation of capillary GC separations, including thermal gradient conditions [69], the comparison of static thermal gradient to isothermal conditions in GC [71] and the comparison of dynamic thermal gradient GC operation to temperature-programmed gas chromatography [71].

Prior to this, the same group of authors had filed a patent on "*Gas chromatography using a thermal gradient that is substantially monotonically non-increasing and has a positive second derivative*" which is presenting two embodiments of the invention claimed to be capable of producing temperature profiles that are monotonically decreasing from injection to detection, or of constant temperature. The distinguishing feature is the fact that with segment-wise created gradients, there would typically be a piece of separation column where, for practical reasons, the temperature profile would increase – in contrast to the present invention [73]. A further patent was filed in 2020 by Tolley and Kingston, aiming at introducing a new realization for both temperature gradient and traveling wave gas chromatography [74]. The patent describes inductive heating of (sections of) a GC column housed in a metal capillary that allows the production of either a monotonically decreasing temperature profile from head to the end of the GC column or to move a heated zone only along the GC column. Although many different forms are presented in this patent which theoretically could produce the desired results, it must be assumed that the idea was never put to work as no chromatograms are presented.

In this context, a further patent deserves mentioning where a "*fluidless column oven for gas chromatography*" is presented in which the GC column (inside a metal capillary) is resistively heated to the desired temperature or temperature profile [75]. The characteristics of this system are that it has a number (6, in the disclosed setup) of individually heated zones: The initial five heated zones allow to create a monotonically decreasing temperature along the column while the last zone is heated again to higher temperature than the penultimate column segment. It is not detailed by the inventors why such a system should bring an advantage over classical isothermally operated, or thermal gradient/temperature-programmed GC systems that have an essentially monotonous increasing or decreasing temperature profile, and it must be doubted that there actually is an advantage in this particular mode of operation (**Figure 15**).

It shall be mentioned that the inventor of this patent is also involved in the production of a TGGC setup that can be fitted into any commercial GC by using its injector and detector; however, replacing the conventional air bath oven with an assembly consisting of a coiled GC column installed over a number of individually addressable heated zones with an external temperature control unit [76]. In contrast to the invention described in

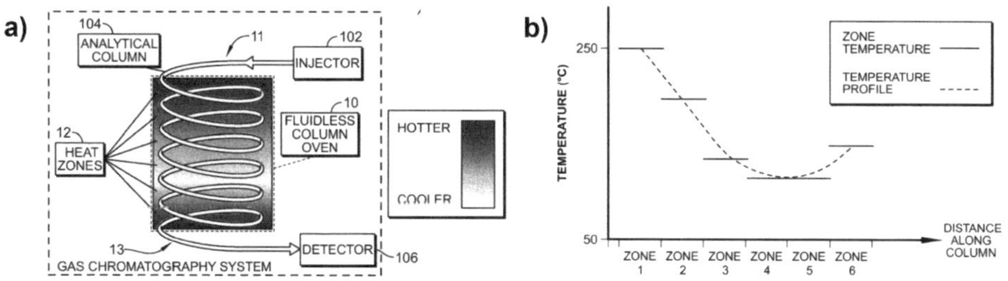

Figure 15.
a) Schematic drawing of the GC system used to create a customized temperature profile with a "fluidless column oven for gas chromatography" and b) resulting non-monotonous temperature profile along the GC column as described in the patent of Pierce [72]). The numbers in the left panel refer to the original patent and denote: 10: Fluidless column oven ('FCO'), 11: Inlet portion, coupled to 102: Injector, 12: Plurality of heat zones, 13: Outlet portion, coupled to 106: Detector, 104: Analytical column (reproduced from DR Pierce, Patent US 10,520,478 B2 (2019). [75]).

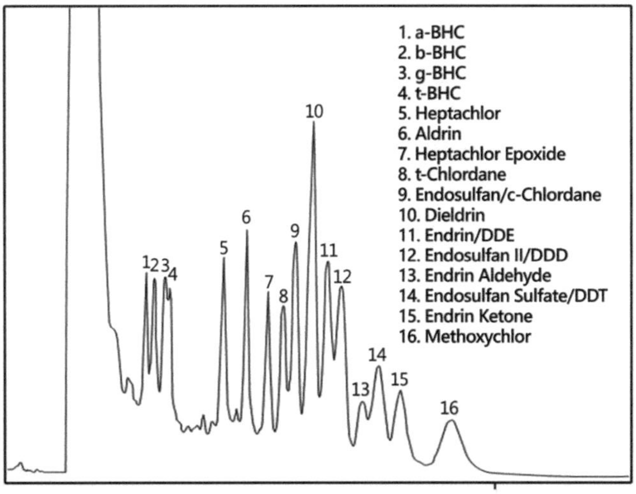

Figure 16.
Separation of a 15 organochlorine pesticides with a GC system with "fluidless column oven", employing the TGGC principle. Column used: Restek MXT-1, 30 m × 0.53 mm, 0.5 µm d_f; inlet: 320°C, outlet: 180°C, FID detection (reproduced from GC Ovens Inc. Website [76]).

the patent, only a monotonous decreasing temperature profile along the column is used, without the ascending final part of the temperature, which is the distinguishing feature of the disclosure in contrast to earlier patents. The proposed system offers the advantage of being able to work with any commercial column of regular dimensions; however, the length of such columns precludes achieving very fast and highly resolved separations, as the optimum heating rate scales with the column length and diameter [15, 77]. Also, as the temperature gradient is not ramped, the late eluting peaks are significantly broadened compared to the early eluting peaks (**Figure 16**).

In contrast to publications and patents that did not lead to a commercial product, the independent development of Boeker at the University of Bonn [78, 79] did lead to a system that eventually also was commercialized [80]. The system consists of a cylindrical tower with a spiral grove from bottom to top along its wall. It is filled with air-permeable foam, open at its bottom end and closed at the top. Centred over the spiral grove, a 1.8 m × 0. 1 mm ID × 0.1 mm d_f GC column is placed inside a stainless steel (SS) capillary that is directly resistively heated. A commercial GC injector and a TOF-MS detector are connected via heated transfer lines. The temperature gradient along the column is formed by operating a ventilator that pushes cold air from the bottom of the cylindrical tower through the foam. Due to the flow resistance, presented to the airflow by the foam inside the cylinder, an airflow gradient is created from bottom to top. At the bottom, the airflow is largest, leading consequently to the strongest cooling of the GC column within the SS capillary, while at the top of the cylinder, the air stream is only faint, leading to much less efficient cooling of the GC capillary. This way, a temperature gradient is created from top where the sample is injected at high column temperature, to the bottom, where the temperature is the lowest at the detector end, which is controlled by the relative strength of heating and cooling (**Figure 17**). The development and characteristics of this system were presented in the initial publication in 2015 [30]. A number of interesting applications were to follow, such as the TGGC/MS separation of explosives [81] or the analysis of residual solvents

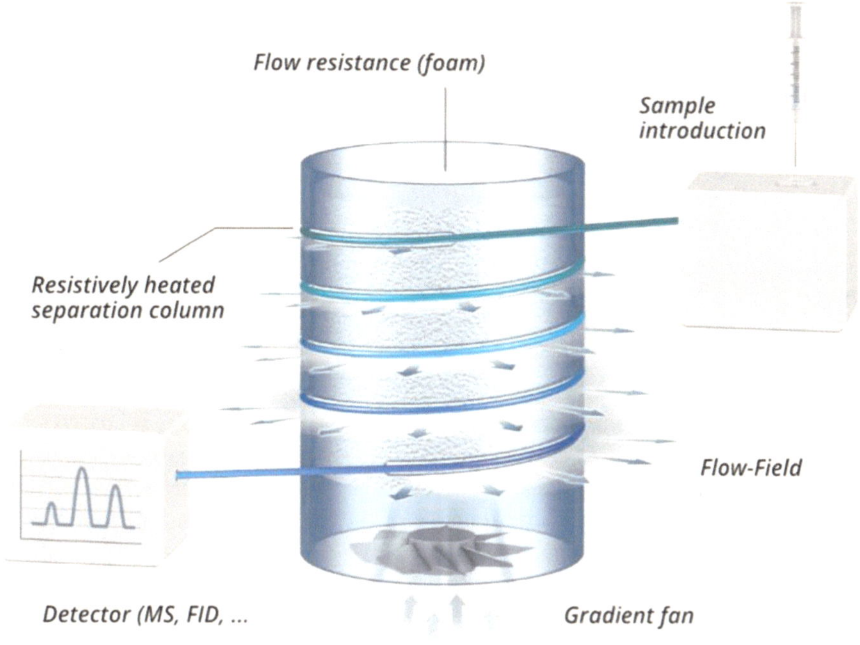

Figure 17.
Schematic representation of the thermal gradient GC system developed by Boeker et al. (Reproduced from the HyperChrom S.A. homepage [80], with permission).

after $CO_{2(l)}$ cryofocusing [82]. Later, Boeker and co-workers, in collaboration with Blumberg, also turned to the theoretical description and modelling of the TGGC separation. Notably the peak profiles and the separation performance of negative thermal gradient operation were discussed in a series of papers [83–85], which can be seen as a scientific dialog to the papers of Tolley and Lee [69–71]. This is particularly so as they were successfully describing both chromatographic separation and peak width.

While the instrument developed by Boeker *et al.* (**Figure 17**) appears to be the only one commercially available that provides maximum performance at dramatically reduced separation time, work is also undertaken in other laboratories to improve the "cooling tower" concept [86], or to develop even more flexible ways of producing thermal gradients [87].

5. Turning theory into viable instrumentation and selected applications of temperature gradient GC

Although the principle of TGGC was already introduced at a very early stage of chromatographic development [36–38], it should take more than six decades until the full potential of this versatile technique is recognized [88]. Much of the delay in appreciating the full versatility of this approach lies in the unavailability of the early landmark papers of Russian authors to the non-Russian speaking community, the scientific correct but in their strict treatment of the matter somewhat apodictic papers of Blumberg *et al.* who pointed out that gradients along the separation column (what

Brumberg called *'nonuniform (coordinate dependent) time-varying separation'*) would not improve chromatographic resolution beyond what is achievable with *uniform time-invariant separation*, e.g. in [34, 89]. However, the biggest challenges that had to be overcome were technical. Much of the ideas that led to the initial prototype of the TGGC system are described in the first column of Boeker [88] which was later extended by a second installment in which he in more detail commented on the technological improvements that allowed the instrument to actually achieve the high performance that it demonstrates today [90].

These improvements relate to the construction of the cooling tower, which in its initial design was a polymer cylinder into which the helical channel was machined and, in the current version of the instrument, is realized with additive manufacturing of the column's support. Using selective laser sintering, internal cooling channels are printed into the wall along the flow channels.

The TGGC module is connected to the injector and the detector via heated transfer lines. This is essential to avoid cold spots, particularly after the separation column, which could be detrimental to the separation already achieved. Moreover, these transfer lines and connectors allow the easy exchange of the separation column (which is to be inserted into the stainless steel tube wound around the supporting structure); however, to adequately fulfill their purpose without adding to peak broadening or distortion, these connectors must be purged. The flow of these connectors must be precisely controlled (*via* electronic pressure controllers, EPCs) to have in the column the flow that is providing optimum separation efficiency. Temperature control becomes of utmost importance, as due to the short column length and separation times, temperature fluctuations in both space and time immediately lead to unstable retention times. The amount of sample injected also is critical: To achieve maximum performance, the column must not be overloaded, which requires high split ratios considering the short length, small ID and low film thickness of the GC columns typically used. This, in turn, requires the use of highly sensitive and fast detectors. Both the FID and time-of-flight mass spectrometers (TOF-MS) are suitable detectors, offering the required sensitivity as well as the necessary data acquisition rate in excess of 100 Hz.

The examples published so far illustrate the advantage of TGGC versus classical modes of operation. These include mainly speed and elution of compounds at lower column temperatures. To illustrate the former advantage, a gas oil sample analysis is reproduced in **Figure 18**. This analysis is completed in 1 minute using a 1.8 m

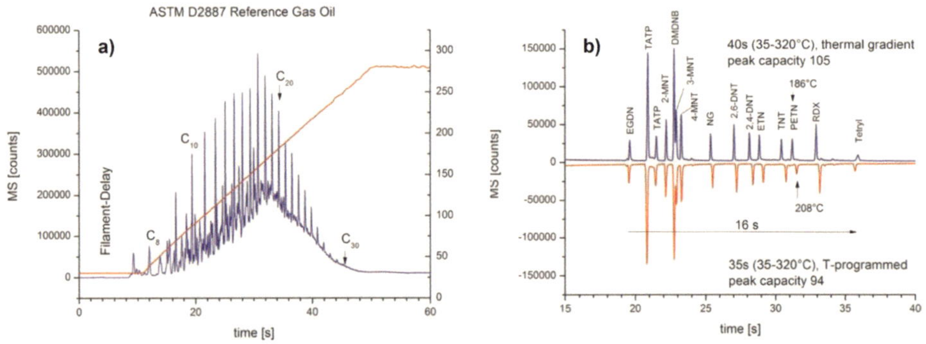

Figure 18.
a) Fast TGGC analysis of an ASTM D2887 reference gas oil sample within 60 s, applying a temperature gradient from 35 to 320°C in 40 sec. b) Analysis of a set of 15 explosives and related substances at two different temperature gradients. (Reproduced from the HyperChrom S.A. homepage [80], with permission).

narrow-bore (0.1 mm ID) column and even offers a better resolution than the standard ASTM method D2887 [91] that proposes a 10 m × 0.52 mm ID wide bore column, leading to a run time of ca. 25 min. Also, the advantage of lower elution temperatures here than in TGGC mode allows eluting even the higher boiling sample constituents below the upper column temperature limit.

This situation has been used to advantage for the analysis of explosive substances which are highly thermolabile. For example, using the somewhat slower gradient (which extends over 40 s), a higher peak is obtained for the most labile substance PETN as compared to the faster gradient (over 35 sec) as the elution temperature of this peak is more than 20°C lower in the case of the slower gradient (**Figure 18**).

6. Conclusions and outlook

The development and (commercial) introduction of TGGC and its beginning acceptance in the scientific community probably represent the greatest innovation in gas chromatography of the last decade, or even after the invention of comprehensive two-dimensional gas chromatography by Liu and Phillips in 1991 [92]. The versatility of this technique to achieve fast, highly resolved separations with short columns is impressive, even if it is accepted by now that the resulting separation cannot be better than the *idealized basic separation* (IBS). However, due to the negative temperature gradient's inherent focusing effect on the analytes, much of the non-ideal behaviour of chromatographic separation can be overcome or reduced, leading to significantly improved peak shape and width.

Practical advantages of lower elution temperature have also been acknowledged, which are equally important for thermally labile analytes and for stationary phases with low upper-temperature limits.

As the GC capillary is directly resistively heated, energy consumption is only a fraction of what an air bath oven GC requires, making this technique more "green" [93].

With the design improvements of the instrumentation that can be expected to benefit, for example, from additive manufacturing [86, 94] or microprocessor control and improved electronics [87], it is expected that TGGC instruments will in the future have an even lower footprint and energy consumption, making them suitable for portable or field-deployable instrumentation.

Making use of the gating ability of a specifically temperature-controlled TGGC setup, it is also anticipated that TGGC will find use in the continuous monitoring of process streams and in comprehensive multidimensional chromatography.

Very likely, TGGC will enable new operational modes of chromatography and their use for advanced applications we may at the current time not even be aware of.

Acknowledgements

The authors would like to gratefully acknowledge the financial support of the work contributing to the development of the TGGC system within the Austrian Research Promotion Agency's project "OPERION" (Proj. No. 879613). Insightful discussions with Dr. Peter Boeker (Univ. of Bonn) are also thankfully acknowledged.

Conflict of interest

The authors declare no conflict of interest.

Acronyms and abbreviations

ASTM	American Society for Testing and Materials
b	proportionality factor
d_f	film thickness of stationary phase
d_c	capillary diameter
$D_{\mathrm{m,i}}$	diffusion coefficient in the mobile phase at inlet condition
$D_{\mathrm{m,o}}$	diffusion coefficient in the mobile phase at outlet condition
D_{s}	diffusion coefficient in the stationary phase
EPC	electronic pressure control (unit)
f_1, f_2	pressure correction factors
f_{gas}	normalized (to column diameter) speed-optimized gas flow rate
FID	flame ionization detector
ΔG	(change in) Gibbs free energy
H	theoretical plate height
ΔH	(change in) enthalpy
IBS	idealized basic separation
ID	inner diameter
ITGC	isothermal gas chromatography
k'	capacity factor (retention factor)
k'_0	capacity factor at infinite dilution
K	partitioning coefficient (distribution coefficient)
L	column length
LP-GC	low-pressure gas chromatography
MDGC	multidimensional gas chromatography
MS	mass spectrometry
N	number of theoretical plates
NTGC	negative thermal gradient gas chromatography
p_{i}	pressure at inlet condition
p_0	pressure at outlet condition
PETN	Pentaerythritol tetranitrate
Q	sample capacity
$\mathcal{R}$	universal gas constant (= 8.314 J mol^{-1} K^{-1})
R_{r}	retention ratio
$R_{\mathrm{T,opt}}$	optimal heating rate
ΔS	(change in) entropy
SOF	speed-optimized gas flow rate
SS	stainless steel
t	retention time
t_r	corrected retention time
t_{m}	dead time
T	absolute temperature
TGGC	temperature gradient gas chromatography
TOF-MS	time-of-flight mass spectrometer
TPGC	temperature-programmed gas chromatography

DOI: http://dx.doi.org/10.5772/intechopen.110591

u_0	mobile phase velocity at outlet condition
V_m	volume of mobile phase
V_s	volume of stationary phase
α	selectivity (selectivity factor)
β	chromatographic phase ratio
β”	solute-liquid phase specific factor
ρ_s	density of the stationary phase.

Author details

Erwin Rosenberg*, Bernhard Klampfl and Robert D. Müller
Vienna University of Technology, Institute of Chemical Technologies and Analytics, Vienna, Austria

*Address all correspondence to: erosen@mail.zserv.tuwien.ac.at

References

[1] McNair HM, Miller JM, Snow NH. Basic Gas Chromatography. 3rd ed. Hoboken, NJ: Wiley; 2019. ISBN-13: 978-1-119-45075-7

[2] Giddings JC. Unified Separation Sciences. Hoboken, NJ: Wiley; 1991. ISBN-13: 978-0-471-52089-4

[3] Wilson RB, Siegler WC, Hoggard JC, Fitz BD, Nadeau JS, Synovec RE. Achieving high peak capacity production for gas chromatography and comprehensive two-dimensional gas chromatography by minimizing off-column peak broadening. Journal of Chromatography A. 2011;**1218**: 3130-3139. DOI: 10.1016/j.chroma.2010.12.108

[4] Klee MS, Blumberg LM. Theoretical and Practical Aspects of Fast Gas Chromatography and Method Translation. Journal of Chromatographic Science. 2002;**40**:234-247. DOI: 10.1093/chromsci/40.5.234

[5] Cramers CA, Janssen HG, van Deursen MM, Leclercq PA. High-speed gas chromatography: An overview of various concepts. Journal of Chromatography A. 1999;**856**: 315-329. DOI: 10.1016/S0021-9673(99)00227-7

[6] van Deursen M, Janssen HG, Beens J, Lipman P, Reinierkens R, Rutten G, et al. Fast Gas Chromatography Using Vacuum Outlet Conditions. Journal of Microcolumn Seperation. 2000;**12**: 613-622. DOI: 10.1002/1520-667X(2000)12:12<613::AID-MCS1007>3.0.CO;2-R

[7] Matisová E, Dömötörová M. Fast gas chromatography and its use in trace analysis. Journal of Chromatography A. 2003;**1000**:199-221. DOI: 10.1016/s0021-9673(03)00310-8

[8] Zoccali M, Tranchida PQ, Mondello L. Fast gas chromatography–mass spectrometry: A review of the last decade. TrAC, Trends in Analytical Chemistry. 2019;**118**:444-452. DOI: 10.1016/j.trac.2019.06.006

[9] Blumberg LM. Theory of gas chromatography. In: Poole CF, editor. Gas chromatography. 2nd ed. Amsterdam: Elsevier; 2021. pp. 19-78. DOI: 10.1016/9,780,128,206,751

[10] Sapozhnikova Y, Lehotay SJ. Review of recent developments and applications in low-pressure (vacuum outlet) gas chromatography. Analytica Chimica Acta. 2015;**899**:13-22. DOI: 10.1016/j.aca.2015.10.003

[11] de Zeeuw J, Peene J, Jansen HG, Lou X. A simple way to speed up separations by GC–MS using short 0.53 mm columns and vacuum outlet conditions. Journal of High Resolution Chromatography. 2000;**23**:677-680. DOI: 10.1002/1521-4168(20001201)23:12<677::AID-JHRC677>3.0.CO;2-L

[12] Giddings JC, Seager SL, Stucki LR, Stewart ŽGH. Plate Height in Gas Chromatography. Analytical Chemistry. 1960;**32**:867-870. DOI: 10.1021/ac60163a043

[13] Wang A, Tolley HD, Lee ML. Gas chromatography using resistive heating technology. Journal of Chromatography A. 2012;**1261**:46-57. DOI: 10.1016/j.chroma.2012.05.021

[14] Dal Nogare S, Bennett CE. Programmed Temperature Gas Chromatography. Analytical Chemistry. 1958;**30**:1157-1158. DOI: doi.org/10.1021/ac60138a600

[15] Blumberg LM, Klee MS. Optimal heating rate in gas chromatography.

Journal of Microcolumn Separation. 2000;**2000**:508-514. DOI: 10.1002/1520-667X(2000)12:9%3C508::AID-MCS5%3E3.0.CO;2-Y

[16] Valentin JR, Hall K, Becker JF. Continuous monitoring of a changing sample by multiplex gas chromatography. Journal of Chromatography. 1990;**518**:199-206. DOI: 10.1016/S0021-9673(01)93175-9

[17] Trapp O. Boosting the throughput of separation techniques by "Multiplexing". Angewandte Chemie, International Edition. 2007;**46**:5609-5613. DOI: 10.1002/anie.200605128

[18] Siegle AF, Trapp O. Development of a straightforward and robust technique to implement hadamard encoded multiplexing to high-performance liquid chromatography. Analytical Chemistry. 2014;**86**:10828-10,833. DOI: 10.1021/ac502933f

[19] Siegle AF, Trapp O. Improving the signal-to-noise ratio in gel permeation chromatography by Hadamard encoding. Journal of Chromatography A. 2016;**1448**:93-97. DOI: 10.1016/j.chroma.2016.04.057

[20] Pallmann S, Siegle AF, Šteflová J, Trapp O. Direct Hadamard transform capillary zone electrophoresis without instrumental modifications. Analytical Chemistry. 2018;**90**:8445-8453. DOI: 10.1021/acs.analchem.8b01010

[21] Blumberg LM, Klee MS. Quantitative comparison of performance of isothermal and temperature-programmed gas chromatography. Journal of Chromatography A. 2001;**933**: 13-26. DOI: 10.1016/S0021-9673(01)01276-6

[22] Giddings JC. Elementary theory of programmed temperature gas chromatography. Journal of Chemical Education. 1962;**39**:569-573. DOI: 10.1021/ed039p569

[23] James AT, Martin AJP. Gas–liquid partition chromatography: The separation and micro-estimation of volatile fatty acids from formic acid to dodecanoic acid. Biochemical Journal. 1952;**52**:242-247. DOI: 10.1042/bj0500679

[24] James AT, Martin AJP, Smith GH. Gas–liquid partition Chromatography. Separation and Microestimation of Ammonia and the Methylamines. Biochemical Journal. 1952;**52**:238-242. DOI: 10.1042/bj0520238

[25] Griffiths J, James D, Phillips C. Adsorption and partition methods. Analyst. 1952;**77**:897-904. DOI: 10.1039/AN9527700897

[26] Dal Nogare S, Harden JC. Programmed temperature gas chromatography apparatus. Analytical Chemistry. 1959;**31**:1829-1832. DOI: 10.1021/ac60155a043

[27] Dal Nogare S, Langlois WE. Programmed temperature gas chromatography. Analytical Chemistry. 1960;**32**:767-770. DOI: 10.1021/ac60163a009

[28] Grob RL, Barry EF. Modern Practice of Gas Chromatography. 4th ed. Hoboken, NJ: Wiley & Sons; 2004. ISBN-13: 978-0-471-22983-4

[29] Rubey WA. Operational theory and instrumental implementation of the thermal gradient programmed gas chromatography (TGPGC) mode of analysis. Journal of High Resolution Chromatography. 1992;**15**:795-799. DOI: 10.1002/jhrc.1240151205

[30] Boeker P, Leppert J. Flow field thermal gradient gas chromatography.

Analytical Chemistry. 2015;**87**: 9033-9041. DOI: 10.1021/acs.analchem.5b02227

[31] Phillips JB, Jain V. On-column temperature programming in gas chromatography using temperature gradients along the capillary column. Journal of Chromatographic Science. 1995;**33**:541-550. DOI: 10.1093/chromsci/33.10.541

[32] Jain V, Phillips JB. High-speed gas chromatography using simultaneous temperature gradients in both time and distance along narrow-bore capillary columns. Journal of Chromatographic Science. 1995;**33**:601-605. DOI: 10.1093/chromsci/33.11.601

[33] Blumberg LM, Jain V, Phillips JB. Focusing cannot enhance resolution or speed limit of a GC column. Journal of Chromatographic Science. 1997;**35**: 451-457. DOI: 10.1093/chromsci/35.9.451

[34] Blumberg LM. Limits of resolution and speed of analysis in linear chromatography with and without focusing. Chromatographia. 1994;**39**: 719-728. DOI: 10.1007/BF02274589

[35] Leppert J, Blumberg LM, Wüst M, Boeker P. Simulation of the effects of negative thermal gradients on separation performance of gas chromatography. Journal of Chromatography A. 1640; **2021**:461943. DOI: 10.1016/j.chroma.2021.461943

[36] Contreras JA. Axial Temperature Gradients in Gas Chromatography. Provo, UT: Brigham Young University, 2010 . https://scholarsarchive.byu.edu/etd/2645. [Accessed: 2022-12-30]

[37] Ettre LS. Chapters in the Evolution of Chromatography. 1st ed. London: Imperial College Press; 2008

[38] Zhukhovitskii AA, Zolotareva OV, Sokolov VA, Turkeltaub NM. New method of chromatographic analysis. Dokl. Akad. Nauk SSSR = Accounts of the Academy of Science of the USSR. 1951;**77**:435-438

[39] Ohline RW, DeFord DD. Chromathermography, the application of moving thermal gradients to gas liquid partition chromatography. Analytical Chemistry. 1963;**35**:227-234. DOI: 10.1021/ac60195a032

[40] Zhukhovitskii AA. Some developments in gas chromatography in the U.S.S.R. In: Scott RPW, editor. Gas Chromatography. Edinburgh: Butterworths; 1960. pp. 293-300

[41] Berezkin VG, Chernysheva TY, Buzayev VV, Ko-shevnik MA. Temperature gradients in gas chromatography. Journal of Chromatography A. 1986;**373**: 21-44. DOI: 10.1016/S0021-9673(00)80206-X

[42] Ettre LS, Zhukhovitskii AA, Berezkin VG. A Russian Pioneer of Gas Chromatography. LC GC North America. 2000;**18**:1148-1155

[43] Moshinskaya MB, Vigdergauz MS. The evolution of the construction and manufacturing of gas chromatographs in the Soviet Union. Journal of Chromatographic Science. 1978;**16**: 351-357. DOI: 10.1093/chromsci/16.8.351

[44] Starshov IM. Analysis of the gases produced by pyrolytic cracking with a HT-2 M cromathermograph. Akad. Nauk SSSR. 1964;**1964**:470-473

[45] Tudge AP. Studies in chromatographic transport III. Chromathermography. Canadian Journal of Physics. 1961;**40**:557-572

[46] Nerheim AG. Gas–liquid chromathermography. Analytical Chemistry. 1960;**32**:436-437. DOI: 10.1021/ac60159a039

[47] Fatscher M et al. Method using a [Chromatography] in which the column temperature is constant with time and varies exponentially along the longitudinal Abscissa. Sciences Chimiques. 1971;**273**:1042-1046

[48] Fatscher M, Vergnaud JM. Use in quantitative analysis of gas chromatography methods with established longitudinal temperature gradient with or without temperature programming. Analusis. 1972;**1**:231-233

[49] Guermouche MH, Fatscher M, Vergnaud JM. Étude des paramètres de la chromatographie en phase gazeuse realisée en couplant un gradient longitudinal et une programmation de température. Journal of Chromatography. 1970;**52**:9-20. DOI: 10.1016/s0021-9673(01)96539-2

[50] Coudert M, Larrat J, Vergnaud JM. Chromatographie en phase gazeuse réalisée simultanément avec une programmation de température et une programmation du gradient longitudinal négatif de température: Théorie de la rétention et influence des paramètres. Journal of Chromatography. 1971;**58**: 159-167. DOI: 10.1016/S0021-9673(00) 96608-1

[51] Fenimore DC. Gradient temperature programming of short capillary columns. Journal of Chromatography. 1975;**112**: 219-227. DOI: 10.1016/S0021-9673(00) 99955-2

[52] Badger CMA, Harris JA, Scott KF, Walker MI, Phillips CSG. Heater-displacement chromatography. Journal of Chromatography. 1976;**126**:11-18. DOI: 10.1016/s0021-9673(01)84059-0

[53] Horrocks JP, Harris JA, Phillips CSG, Scott KF. Quantitative aspects of heater displacement chromatography. Journal of Chromatography. 1980;**197**:109-119. DOI: 10.1016/s0021-9673(00)81227-3

[54] Rubey W. Gas chromatography methods and apparatus. US Patent 4,923,486. 1990

[55] Rubey WA. Gas Chromatography Methods and Apparatus US Patent 5,028,243. 1991

[56] Rubey WA. A different operational mode for addressing the general elution problem in rapid analysis gas chromatography. Journal of High Resolution Chromatography. 1991;**14**: 542-548. DOI: 10.1002/jhrc.1240140806

[57] Jain V, Phillips JB. Fast temperature programming on fused-silica open-tubular capillary columns by direct resistive heating. Journal of Chromatographic Science. 1995;**33**:55-59. DOI: 10.1093/chromsci/33.1.55

[58] Phillips JB, Jain V. On-column temperature programming in gas chromatography using temperature gradients along the capillary column. Journal of Chromatography Science. 1995;**33**:541-550. DOI: 10.1093/chromsci/ 33.10.541

[59] Jain V, Phillips JB. High-speed gas chromatography using simultaneous temperature gradient in both time and distance along narrow-bore capillary columns. Journal of Chromatographic Science. 1995;**33**:601-605. DOI: 10.1093/ chromsci/33.11.601

[60] Hiller JF, Hughes GH, Martin DP. Apparatus and method for establishing a temperature gradient in a chromatography column. US Patent 5,215,556. 1993

[61] Rounbehler DP, Achter EK, Fine DH, Jarvis GB, MacDonald SJ, Wheeler DB, Wood CD. High speed gas chromatography. US Patent 5,808,178. 1998

[62] Zhao H, Yu L, Zhang J, Guan Y. Characteristics of TGPGC on short micro packed capillary column. Analytical Sciences. 2002;**18**:93-95. DOI: 10.2116/analsci.18.93

[63] Contreras JA, Design and Application of Thermal Gradient Programming Techniques for Use in Multidimensional Gas Chromatography–Mass Spectrometry (MDGC-MS). Dayton, OH: University of Dayton, 2004. Available from: https://ecommons.udayton.edu/graduate_theses/2103 [Accessed: 2022-12-30]

[64] Contreras JA, Rockwood AL, Tolley HD, Lee ML. Peak sweeping and gating using thermal gradient gas chromatography. Journal of Chromatography A. 2013;**1278**:160-165. DOI: 10.1016/j.chroma.2013.01.010

[65] Contreras JA, Wang A, Rockwood AL, Tolley HD, Lee ML. Dynamic thermal gradient gas chromatography. Journal of Chromatography A. 2013;**1302**:143-151. DOI: 10.1016/j.chroma.2013.06.008

[66] Wang A. Microchip Thermal Gradient Gas Chromatography. Provo, UT: Brigham Young University, 2014. Available from: https://scholarsarchive.byu.edu/etd/4300. [Accessed: 2022-12-30]

[67] Wang A, Hynynen S, Hawkins AR, Tolley SE, Tolley HD, Lee ML. Axial thermal gradients in microchip gas chromatography. Journal of Chromatography A. 2014;**1374**:216-223. DOI: 10.1016/j.chroma.2014.11.035

[68] Tolley HD, Tolley SE, Wang A, Lee ML. Moving thermal gradients in gas chromatography. Journal of Chromatography A. 2014;**1374**:189-198. DOI: 10.1016/j.chroma.2014.10.090

[69] Tolley HD, Avila S, Iverson BD, Hakins AR, Porter NL, Johnson SL, et al. Simulating capillary gas chromatographic separations including thermal gradient conditions. Analytical Chemistry. 2021;**93**:2291-2298. DOI: 10.1021/acs.analchem.1c00438

[70] Avila S, Tolley HD, Iverson BD, Hakins AR, Porter NL, Johnson SL, et al. Comparison of static thermal gradient to isothermal conditions in gas chromatography using a stochastic transport model. Analytical Chemistry. 2021;**93**:6739-6745. DOI: 10.1021/acs.analchem.1c00438

[71] Avila S, Tolley HD, Iverson BD, Hakins AR, Johnson SL, Lee ML. Comparison of the dynamic thermal gradient to temperature-programmed conditions in gas chromatography using a stochastic transport model. Analytical Chemistry. 2021;**93**:11785-11,791. DOI: 10.1021/acs.analchem.1c02210

[72] Avila S. Effects of Static and Dynamic Thermal Gradients in Gas Chromatography. Provo, UT: Brigham Young University: 2021. Available from: https://scholarsarchive.byu.edu/etd/8778. [Accessed: 2022-12-30]

[73] Tolley HD, Wang A, Tolley SE, Lee ML, Hawkins AR. Gas chromatography using a thermal gradient that is substantially monotonically non-increasing and has a positive second derivative. Patent US 9,678,046B2. 2017

[74] Tolley S, Kingston C. Heaters and Their Use in Temperature Gradient and

Traveling Wave Chromatography. Patent US 2021/0208115 A1. 2021

[75] Pierce DR. Fluidless column oven for gas chromatography system. Patent US 9,927,406 B1 (2018), also: US 10,520,478 B2. 2019

[76] GC Ovens Inc., Henderson, NV, see: www.gcovens.com [Accessed: 2023-01-10]

[77] Klee MS, Too Fast GC. What is too Fast in GC? Separation Science blog. Available online on: https://blog.sepscience.com/gaschromatography/too-fast-gc-what-is-too-fast-in-gc [Accessed: 2023-01-10]

[78] Boeker P. Strömungsfeld induzierte Temperatur-Gradienten-Gaschromatographie. Patent DE 102014 004286 B3 (2015.04.16)

[79] Boeker P. Flow-Field-induced temperature gradient gas chromatography. Patent US 10,641,748 B2 (2020)

[80] HyperChrom S.A. https://www.hyperchrom.com/ [Accessed: 2023-01-10]

[81] Leppert J, Härtel M, Klapötke TM, Boeker P. Hyperfast flow-field thermal gradient GC/MS of explosives with reduced elution temperatures. Analytical Chemistry. 2018;**90**:8404-8411. DOI: 10.1021/acs.analchem.8b00900

[82] Chopra MD, Müller PJ, Leppert J, Wüst M, Boeker P. Residual solvent analysis with hyper-fast gas chromatography–mass spectrometry and a liquid carbon dioxide cryofocusing in less than 90 s. Journal of Chromatography A. 1648;**2021**:462179. DOI: 10.1016/j.chroma.2021.462179

[83] Leppert J, Blumberg LM, Boeker P. Equation for evolution of temporal width of a solute band migrating in chromatographic column. Journal of Chromatography A. 1612;**2020**:460645. DOI: 10.1016/j.chroma.2019.460645

[84] Leppert J, Müller PJ, Chopra MD, Blumberg LM, Boeker P. Simulation of spatial thermal gradient gas chromatography. Journal of Chromatography A. 2020;**1620**:460985. DOI: 10.1016/j.chroma.2020.460985

[85] Leppert J, Blumberg LM, Wüst M, Boeker P. Simulation of the effects of negative thermal gradients on separation performance of gas chromatography. Journal of Chromatography A. 1640; **2021**:461943. DOI: 10.1016/j.chroma.2021.461943

[86] Müller RD, Design and Construction of a Thermal Gradient Module for Gas Chromatography, Vienna: Vienna University of Technology. 2021. Available from: https://doi.org/10.34726/hss.2021.85346 [Accessed: 2022-12-20]

[87] Klampfl B. Development of a Negative-Thermal Gradient GC System for Fast Gas Chromatography and Its Application to the Study of LIB Volatile Products Formation. Vienna: Vienna University of Technology; 2023

[88] Boeker P. Beginner's Luck and Hyper-Fast GC. The Analytical Scientist. 2016;**09**(16):38-42. Available from: https://theanalyticalscientist.com/techniques-tools/beginners-luck-and-hyper-fast-gc [Accessed: 2023-01-10]

[89] Blumberg LM. Outline of a theory of focusing in linear chromatography. Analytical Chemistry. 1992;**64**: 2459-2460. DOI: 10.1021/ac00044a028

[90] Boeker P. Hyper-fast GC – and a new world perspective. The Analytical Scientist. 2019;**05**(19):30-35. Available

from: https://theanalyticalscientist.com/techniques-tools/hyper-fast-gc [Accessed: 2023-01-10]

[91] Standard Test Method for Boiling Range Distribution of Petroleum Fractions by Gas Chromatography; ASTM D2887-14; ASTM International: West Conshohocken, PA, 2014

[92] Liu Z, Phillips JB. Comprehensive two-dimensional gas chromatography using an on-column thermal desorption modulator interface. Journal of Chromatographic Science. 1991;**29**: 227-231. DOI: 10.1093/chromsci/29.6.227

[93] Aly AA, Górecki T. Green chromatography and related techniques. In: Płotka-Wasylka J, Namieśnik J, editors. Green Analytical Chemistry. Green Chemistry and Sustainable Technology. Singapore: Springer; 2019. pp. 241-298. DOI: 10.1007/978-981-13-9105-7_9

[94] Gross B, Lockwood SY, Spence DM. Recent advances in analytical chemistry by 3D printing. Analytical Chemistry. 2017;**89**:57-70. DOI: 10.1021/acs.analchem.6b04344

Chapter 3

Uses of Portable Gas Chromatography Mass Spectrometers

Robert Owen Bussey III

Abstract

Gas chromatography mass spectrometry (GCMS) technology, whether in a laboratory or in the field allows scientists to identify and quantitate volatile and semi-volatile chemical compounds at low levels. It was not until the 1990s, well after the birth of GCMS in the 1950's, that portable GCMS technology became possible. GCMS miniaturization along with a need for scientists to test samples outside of the laboratory drove the development of portable GCMS systems. Currently, scientists in the environmental, emergency response, government, military sectors, and private manufacturing industries use portable GCMS technology to monitor and quantitate various chemicals such as low levels of hazardous compound exposure in the environment. Successful implementation of portable GCMS also required that many sample preparatory techniques used in the laboratory must be modified for application in the field to maintain simplicity and robustness of the analysis of complex matrices like soil or water. This chapter will describe portable GCMS technology along with the current uses and sample preparatory techniques utilized.

Keywords: portable gas chromatography, mass spectrometry, sample preparation, field deployment, GCMS technology

1. Introduction

Portable gas chromatography mass spectrometers (GCMS) have been used by people in the environmental science, emergency response, government, military, and private manufacturing sectors [1]. Common analysis targets in the field have been narcotics, explosives, hazardous industrial environmental contaminants, and even food/beverage biomarkers for counterfeit prevention. These systems can identify and quantitate volatile and semi-volatile compounds contained in gases, liquids, and solids [1]. The sensitivity of this technique is high with the limit of detection for some chemical targets ranging from parts-per-billion to parts-per-trillion.

This section gives an overview of the topics reviewed in this chapter including portable GCMS technology, user requirements, and method development. Section 2 contains information about the basic technology behind portable GCMS systems. This includes information on low thermal mass gas chromatography column technology. It also includes key technology differences including available sample

 IntechOpen

preparation devices for the following commercial portable GCMS systems: Perkin Elmer Torion T-9series, FLIR Griffin 500 series, and Infincon Hapsite ER series. All systems have different, yet similar characteristics that need to be evaluated by users to establish the best system for their specific methods and deployment criteria. Section 3 will provide specific case studies highlighting the different systems and techniques.

Section 3.1 will give characteristics important to users. The users for a Homeland Security study designed testing parameters to evaluate different portable GCMS technology specifically to see which characteristics were most important for first responders or investigators in the field. The study suggested that many different parameters were important, but again each user would need to evaluate which parameters fit better with their specific application.

Section 3.2 contains a method development review for headspace needle trap quantitation of butylated hydroxy toluene using a portable GCMS. The method will describe the effort involved in creating a purge and trap method for the analyte of interest and comparison with established GCMS preparatory methods. This work allowed for better quality control of cosmetics containing butylated hydroxy toluene using portable GCMS technology.

Section 3.3 contains a method development review for measuring 24 different illicit drugs with adulterants which could interfere with quantitation using portable GCMS technology. Method development was needed to show that the results created with the portable GCMS technology matched benchtop GCMS technology in a laboratory.

Section 3.4 showed that by using focusing agents for thermal desorption some portable GCMS technology had better precision when measuring chemical weapons agents. These focusing agents were necessary to decrease the relative standard deviation in order to show good chemical weapon agent quantitation reproducibility in the field using portable GCMS technology. These study descriptions show the amount of data required to create reliable quantitation methods for portable GCMS technology out the field.

2. Characteristic portable GCMS technology

Different commercialized portable GCMS technology have unique features overlaying a set of common core elements. In this next section, the technology that makes portable GCMS distinct from laboratory based GCMS technology will be discussed in detail to show why this technology can be deployed outside of the laboratory. In addition, vendor specific sample preparatory technology will be discussed. Review of the product brochures and vendor specific documentation are included to describe the following portable GCMS systems: Perkin Elmer Torion T-9, FLIR Griffin G510, and Hapsite ER systems. These three instruments are the most up to date examples of portable GCMS systems in the literature.

2.1 Low thermal mass gas chromatography column technology

Besides the GCMS column selectivity, the temperature programming is one of the most important factors in contributing to a successful GCMS separation. Temperature programming is a process of increasing and decreasing column temperature during a GC run [2]. The temperature changes over a GC run contribute to the separation of compounds with different boiling points, and can be used to change detection limits, help with peak symmetry, and facilitate column cleaning of high boiling point compounds [2].

For normal GC systems, a convection oven requiring 220 volts is used to heat a silica capillary column to achieve the programmed temperature control mentioned above. The heating of normal GCMS ovens requires a lot of energy and is relatively fast. The cool down time of the oven can be much longer as the oven approaches room temperature since it is only cooled by a fan. All the materials that make up the oven store the heat which in turn need to be cooled.

The temperature programming of a portable GCMS column allows for faster changes in temperature rates for heating and cooling and has roughly 1% of the energy usage compared to a laboratory based GCMS system [3]. The technology used to achieve this is called low thermal mass (LTM) GC. Unlike a convection oven of a normal GC system, the LTM GC technology combines silica column, heating elements, insulation, and temperature sensors in one assembly as to require less material to heat up and cool down. Less required energy means less time required to heat up and cool LTM GC columns [3]. This is important for GCMS miniaturization for portable systems to conserve energy on battery power and to maintain the ability to run with a column up to 30 meters in length.

The LTM GC column assemblies consist of the following parts: insulated silica capillary column, insulated heating wire, resistive temperature detector (as a temperature sensor), and metal foil. Each component is bundled together and then twisted into a torus shape formed by rotating the closed loop of bundled components to create a smaller loop. **Figure 1** shows the composition of the LTM GC column torus configuration along with the cross section of the bundle. This configuration allows for a high surface area and thus rapid heating and cooling potential while maintaining low power consumption due to the fact there is no large GC oven, just the heating wires in the torus bundle. The absence of a large GC oven means the overall heat capacity of the system is smaller and thus the amount of energy required to heat or cool is much smaller. Faster heating and cooling rates allow for smaller runtimes and potentially better peak resolution depending on specific methodology required.

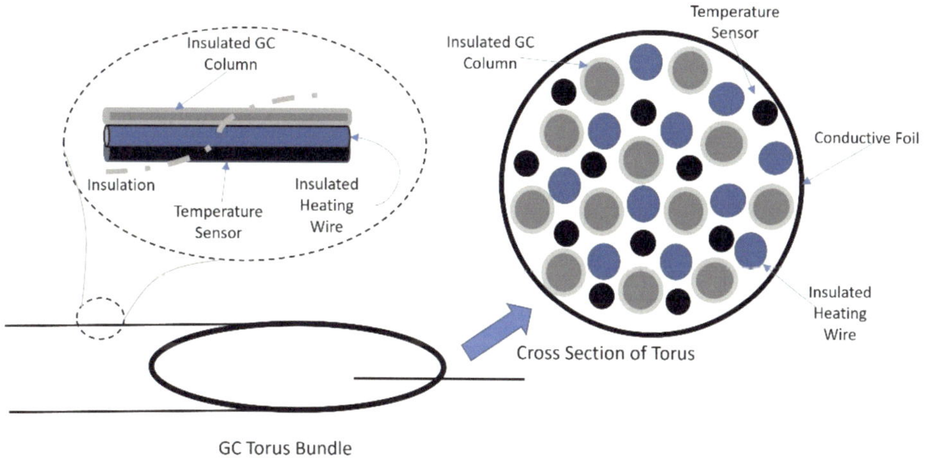

Figure 1.
A diagram of the LTMGC assembly and a cross section of the bundle.

2.2 Commercialized portable GCMS systems

As discussed above, the LTM GC column set up is shared among many of the portable GCMS systems. In this section, three of the following commercialized portable GCMS systems will be discussed from literature available from the manufacturer: Perkin Elmer Torion T-9series, FLIR Griffin 500 series, and Infincon Hapsite ER series. Method development required to create methods in the field will be discussed later in the Section 3 case studies. As of the writing of this chapter, these three instruments are the most up to date examples of portable GCMS systems in the literature. In addition, the vendor-specific sample preparation technology associated with each portable GCMS technology will be discussed.

2.2.1 Perkin Elmer Torion T-9 GCMS product brochure information

The weight of the Torion T-9 portable GCMS is 32 pounds. Battery power is up to 2.5 hours. This system has the option of an internal disposable helium cylinder or an external hookup for a larger helium container. From a cold start, the system is ready to run samples within 5 minutes. Automated startup performance validation is done to check system performance [4]. The LTM GC column allows for the analysis of volatile organic compounds (VOCs) and semi-volatile organic compounds (SVOCs) with the ability to match performance from lab based GCMS systems. Small diameter GC columns along with rapid temperature programming allows for shorter GC analysis times. The maximum temperature ramp rate is 2.5°C per minute.

The Toroidal ion trap allows for large trapping volumes which can increase signal to noise and spectral quality with a scan range of 41–500 m/z [4]. The color touchscreen allows for easy changes in the method and for data analysis. On-board automated Chromion® software/option PC integration allow for the potential of compound deconvolution and identification with the National Institute of Standards and Technology (NIST) libraries [4].

Some of the Perkin Elmer sample preparation technologies for portable GCMS systems technologies are as follows for Custodian® Sampling Devices: Solid Phase Micro Extraction (SPME), Coil Microextraction (CME), and Needle Trap (NT). The Custodian® devices are made of hard plastic and have a push button trigger that allows certain parts of the extraction devices to exit and retract into a 19-gauge needle [5]. The SPME device exposes a SPME fiber PDMS / DVB 65 μm to concentrate volatile compounds when exposed to sealed headspace or direct immersion into water/ liquid samples [5]. The CME device exposes a steel coil that draws up dissolved solid and liquid samples [5]. The solvent is then allowed to evaporate before it is retracted into the 19-guage needle and is often best for SVOCs [5, 6]. The NT device contains a protected tribed Tenax TA, Carboxen 1016 and Carboxen 1003 fiber [5]. This fiber can be exposed to liquid samples when coupled thermal desorption units or a pump can pump air through it with a purge and trap apparatus for air samples [5]. All fiber sorbents or coils that have been retracted into the Custodian® device can be inserted into the heated GCMS inlet for volatilization according to specific parameters of the matrix holding the chemical compounds.

2.2.2 FLIR Griffin 500 GCMS

The weight of the FLIR Griffin 500 portable GCMS is 36 pounds. Battery power is 2 to 4 hours depending on the scanning mode. This system also has the option of an internal disposable helium cylinder or an external hookup for a larger helium container. From a cold start, the system is ready to run samples within 15 minutes. This system also uses a LTM GC column assembly for the analysis of VOCs and SVOCs with the ability to match performance from lab based GCMS systems. The mass spectrometer in this system is a linear quadrupole that allows scanning from 15 to 515 m/z [7]. The color touchscreen allows for any changes in the method and for data analysis. On-board software allows for the potential of compound deconvolution and identification with the NIST and Scientific Working Group for the Analysis of Seized Drugs (SWGDRUG) libraries [7].

The Griffin portable GCMS systems have many different sample introduction techniques which include air, liquid, and solid samples. There is an integrated heated sample probe that can intake air sample from the environment during survey mode [7]. Split/splitless injection ports are present for different forms of sample introduction including air, liquid and SPME fibers. Syringes can be used to inject gas or liquid samples into the GCMS [8]. SPME fibers can be either used in sealed headspace experiments or immersed in liquids like water or solvents to concentrate the chemical compounds. Any commercial SPME fiber can be used and then put into the Griffin GCMS injection port. External headspace heating devices are available for purchase. When many liquid samples need to be injected into the GCMS with reproducible volume measurements, Griffin offers an autosampler that can hold up to 120 samples. Another sample prep device is the PSI-Probe. It is used with the Touch-and-Go (TAG) technology to collect solid or liquid samples just by touching the TAG sampler to the sample and then dropping it into the probe for thermal desorption. No solvents or sample prep is needed according to FLIR [8]. External headspace heating devices are also available for purchase. In addition, GERSTEL-Twister bars, spin plate bars coated in sorbent, can be used to extract compounds from liquid samples. The bars are said to be more sensitive than SPME and can be dropped into the PSI-probe once they are dry [8].

2.2.3 Infincon Hapsite ER portable GCMS system

The weight of the Inficon Hapsite ER portable GCMS is 42 pounds. Battery power is between 2 to 3 hours. This system has the option of an internal disposable nitrogen cylinder or an external hookup for a larger nitrogen container. Tuning occurs with an internal gas cylinder containing an internal standard [9]. This system also uses a LTM GC column assembly for the analysis of VOCs and SVOCs with the ability to match performance from lab based GCMS systems. The mass spectrometer allows scanning from 41 to 300 m/z [9]. The touchscreen allows for any changes in the method and for data analysis. On-board software does compound data processing with the National Institute for occupational Safety and Health (NIOSH) database. The NIST library is also available on a laptop for processing [9].

The Inficon Hapsite ER system has many options to introduce samples into the system. The air probe introduces volatiles compounds directly into the portable GC system [8]. In addition, the air probe can be hooked up to other available accessories. The headspace sampling system is a battery-operated accessory that can heat liquid and solid samples in vials, create headspace volatiles, and then the headspace is pumped directly into the Hapsite system with the air sampling probe [9]. The Thermal Desorber Sampling System can be attached directly to the Hapsite ER universal interface to allow suction of air for defined times through a TDU sorbent tube and then subsequent desorption of that sorbent into the mass spec for analysis [8]. The SPME sampling System can also be attached to the Hapsite ER universal interface for SPME fiber introduction which can extract samples from gas headspace or liquid samples [9]. Also, the Situ Probe Purge and Trap GCMS is a battery-operated accessory allowing for purge and trap sampling of water head space samples directly to the Hapsite ER system through the air sampling probe [8].

2.3 Commercialized portable GCMS conclusions

There are some characteristics of commercialized portable GCMS systems that were not included in the discussion. The previous sections showed that portable GCMS systems have some similar features such as LTM GC columns and sample preparation techniques. Each manufacturer has different variants of how to perform SPME and headspace. Overall, every user will have to evaluate each characteristic and potentially request a demo model to evaluate which system is best for their required method and their required deployment area such as a war zone, manufacturing facility or storage facility for illicit drugs.

3. What aspects of portable GCMS systems are useful to the user?

3.1 Important considerations for portable GCMS users in the field

What are the important considerations relevant to using portable GCMS in the field according to the experts? Many first responders, scientists, and other portable GCMS users have shared what portable GCMS characteristics are important to them as part of the System Assessment and Validation for Emergency Responders (SAVER) program which is under the National Urban Security Technology Laboratory (NUSTL) and the U.S. Department of Homeland Security (DHS).

The SAVER program gathers product information on commercially available equipment, conducts impartial testing led by experts in the first responder community, and then provides the results to allow for better informed decision-making when users buy equipment [1].

The SAVER program assessed several commercially available portable GCMS systems in order to demonstrate the important characteristics required by expert users in the field. The case studies in Sections 3.1.1 and 3.1.2 describe the study parameters used by the SAVER program to determine the characteristics important to users. The experts used in this study each had 8 to 20+ years in firefighting, public health, law enforcement, national guard weapons of mass destruction, civil support, and hazardous materials management. This information was then published for users to make informed decisions on capabilities of portable GCMS technology [1]. This section is provided to familiarize the reader with general portable GCMS function and what functions are important to GCMS users in the field and outside of the controlled laboratory environment. Different samples were analyzed to assess the following portable GCMS systems for usability, ease of deployment, maintainability, and capability: Griffin G-510, Torion T-9, and Hapsite ER. Sample preparation techniques/technology will be discussed in more detail in Section 3 as it pertains to method development.

3.1.1 Unknown analysis of portable GCMS systems

Each portable GCMS system was assessed for the ability to analyze for volatile or semi-volatile compounds in vapor from a sample. This test analyzed the following consumer products: isopropyl alcohol, ethyl acetate-based, and acetone-based nail polish remover [1]. All of these products contained volatile organic compounds with large vapor pressures in simplistic matrices for preliminary testing.

Each instrument had different ways of introducing the samples into the system. The testers used vapor sampling probes for the Griffin G-150 and Hapsite ER systems and a Solid Phase Micro Extraction (SPME) fiber collection device for the Torion T-9 to introduce the samples into the GCMS systems.

All the systems were able to identify isopropyl alcohol, ethyl acetate, and acetone in each of the consumer products along with relative chemical compositions listed by the manufacturers [1]. The evaluation then moved onto the second stage of chemical composition testing.

3.1.2 Liquid and solid sample testing

The next round included each evaluator testing at least one liquid and one solid sample. The liquid samples available for testing were fabric spot cleaner, mentholated electronic cigarette liquid, caffeinated beverage, oral analgesic spray lubricant, and liniment. The solid samples available for testing included wintergreen candy, wood filler, ibuprofen, aspirin pill, and caffeine pill [1]. These solid and liquid samples were included not only for the complex matrix, but also because of the semi-volatile/higher molecular weight ingredients. These compounds required longer analysis time compared to the volatile compounds in the previous experiments due to higher molecular weight/semi-volatile compounds having physical properties which cause longer elution times on the GC columns.

The evaluators repeated the use of the sampling probes and the SPME fibers from the previous section to extract any volatile compounds from the liquid and solid samples. Also, the solid and liquid samples were dissolved and diluted in organic solvent. These samples were then injected into the sample injection port of the Griffin G-150 with a calibrated GC syringe accounting for specific volumes to measure the analyte concentration. The samples dissolved in organic solvent were also extracted by submerging a Coiled Microextraction (CME) device manufactured by PerkinElmer. This CME device was then injected into the Torion T-9 using the instrument's sample injection port.

Additional samples were also analyzed due to availability in their immediate environment. The sample vapor probes for the Griffin G-150 and Hapsite ER instruments were used to collect emissions from vehicles in a parking lot. The same vapors were also sampled with a SPME fiber and then inserted into the Torion T-9 for a full GCMS analysis. The volatile organic compounds emitted by pine trees next to the parking lot were sampled with the sample probes of the G-150 and Hapsite ER. Sampling occurred either next to the pine needles or at the top of a vial containing harvested pine needles. In addition, a SPME fiber was used to collect the volatile pine needle compounds from the headspace of a vial and then the SPME fiber was put into the Torion T-9. Another sampling technique for the Torion T-9 included a battery-operated air sampler. This device sucked air/ vapor close to the branch of a pine tree. This air sampler contained a detachable needle trap device (NTD) with sorbent. This device was then detached and inserted into the Torion T-9. Samples were analyzed with full scan GCMS.

The chemical composition of each analysis contained compounds that matched reported chemical compositions in the literature. In addition, to evaluate accessibility and usability while wearing personal protective equipment, experts used Level A gloves from encapsulated suits commonly used to completely protect users during hazardous materials/biological/chemical spills.

3.1.3 Report summary about major considerations for portable GCMS acceptability

This evaluation used the testing procedures described above to decide what were the most important characteristics of portable GCMS systems in the following categories: usability, ease of deployment, maintainability, and capability. The scores of each portable GCMS system will not be discussed in this section, but the main characteristics of portable GCMS systems required for successful experiments will be discussed.

The usability or ease of performing analyses in the field increases the chance of successful experiments. Easy to learn interfaces with the built-in touch screen along with intuitive operating systems were important. The touch screen could even be operated while the experts were wearing Level A gloves in a self-encapsulated hazardous materials suit. The software's ability to switch between basic settings and advanced settings would allow the user experience to be varied based on user availability in the field. Remote function of the system would be beneficial if the instrument had to be mounted to a robotic system for remote transport into extremely hazardous conditions. Ease of sampling with sampling probes or cheap/innovative consumables gives the user more flexibility in the field when the matrix may be gas, liquid, or solid. With more use also comes the necessity to count injection to estimate carrier gas usage of the internal helium or nitrogen gas cylinders in the system. Some users suggested internal gas pressure levels to directly measure the gas, an integral part of the system. In addition, the inclusion of multiple gas cylinders within a system would allow the possibility of gas cylinder replacement without having to shut down the instrument. All these observations would increase the chances of successful experiments in the field.

The deployment and maintainability, along with ease of basic setup and startup in the field, were important to the end users. Multiple batteries to power the system was desirable to lengthen the analysis time. In addition, charging capabilities must be as fast as possible with the inclusion of vendor independent batteries for ease and cost of replacement. Operating temperature of the instrument must be able to handle both hot and cold environments. Remote access and diagnosis of problems by vendors was suggested to prevent instrument downtime and detachable/replaceable parts like inlet covers were required to allow for decontamination, if necessary, in toxic environments.

The main instrument capability category or scientific measurement ranges were also evaluated. Large ranges in atomic mass units from small to large were essential for an expert to measure volatile or semi-volatile compounds that could range in size in an unknown matrix. A wide column temperature range would also help in this aspect so that separation from a complex matrix would be possible. Automated data analysis/processing tools used with mass spectral library identification, report generation, and sending reports by email were highly rated in this process.

In summary, many different aspects of portable GCMS systems were evaluated. None of the portable GCMS systems failed the analytical testing performed by the experts. Usability with sample preparation, ease of sample introduction, temperature limits, mass scanning limits, and ability to swap battery and gas canisters were important to the experts. This report gave general opinions about operation and usability parameters important to portable GCMS systems. Each system needs to be evaluated by target users to determine if those parameters are important for their specific applications.

3.2 Case study 1

3.2.1 Case study 1: Introduction

Case study 1 relates to creating a method for measuring butylated hydroxytoluene (BHT) in cosmetics [10]. As seen in **Figure 2**, BHT is one of many antioxidants used in skin care products. Antioxidants are included in these products to help prevent the appearance of dark spots, wrinkles, and changes in skin elasticity. The antioxidants act directly to eliminate free radicals which damage skin DNA, lipids, and proteins which in turn causes the signs of skin aging described above. The free radicals come from exposure to sun and other environmental contaminants [10, 11].

BHT penetrates the skin, acts on free radicals, and then any residual BHT remains within the layers of the skin. It is suggested that long term exposure to the skin can contribute to increased toxicity in various organ systems in the human body [10, 12]. Toxicity issues may be mitigated by decreasing exposure dosage or exposure duration. Despite having potential toxicity issues, products do contain this antioxidant. Methods of measurement are required to serve as quality control tests for manufactures to prevent overages that may be excessive.

A previous GCMS methods available for quantitating BHT included solvent extraction/Solid Phase Extraction (SPE) in water and then evaporation for GCMS analysis [13]. This method was made to be simpler and more robust than other techniques in order to rapidly quantitate BHT in cosmetics in manufacturing facilities using a portable GCMS system. This study chose lotion with BHT as a representative cosmetic since lotions have long exposure times on human skin due to the method of use.

3.2.2 Case study 1: Portable GCMS parameters

The Torion Tridion-9 portable GCMS system was used to analyze for BHT in this study [10]. As previously mentioned, this portable GCMS system can accept Solid Phase Micro Extraction (SPME) fibers or tubes/devices containing sorbent material.

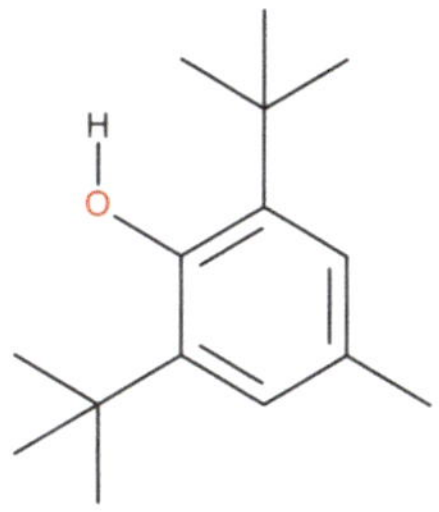

Figure 2.
Butylated hydroxy toluene structure.

Once the objects containing the sorbent material were inserted into the portable GCMS, the analytes of interest bound to the sorbent were desorbed and entered the GCMS for analysis both with split and splitless sample flow. The inlet temperature was 280°C to ensure complete desorption. As the sample was desorbing, the resulting volatized compounds were drawn into the GCMS. The inlet desorption method was the following: 280°C hold for 1 second splitless air flow, 10:1 sample air split flow for 10 seconds, and a 50:1 sample air flow split for 30 seconds. The splitless and split air flow sections were combined to balance higher sensitivity (splitless) with high enough air sample flow rates (split) to desorb samples fast enough to maintain good peak shape of the matrix and BHT [10]. The oven method was as follows: 37°C for 2 seconds, increase to 220°C at a rate of 2°C per second, and then a 50 second hold at 220°C (approximate chromatographic separation of 2.4 minutes). GC ion-trap heater was 155°C and transfer line temperature was 250°C. Calibration and performance validation was done with Perkin Elmer's Calion-13 standard mixture. The portable GCMS mass analyzer was run at 70 eV with an in-trap electron gun source. The GCMS scanned from 43 to 500 Da with a scan time of 50 ms.

3.2.3 Case study 1: SPME headspace method

This study performed several sample preparation techniques to compare results and see if any differences appeared between methods. The first method started with a standard addition calibration curve to quantitate BHT in lotion due to a matrix interference potential affecting the signal. Known quantities of BHT were spiked into the lotion matrix which contained an unknown starting BHT concentration. This method was also repeated with lotion which did not contain BHT. BHT standards of 10, 7, 5, 3, and 1 μg/mL were prepared by diluting a stock with acetone to obtain the desired concentrations. Ten microliters of each standard was spiked onto separate 0.1 g samples of the lotion suspended in 4 mL of water. Each sample was vortexed for 5 minutes and then further mixed with a rotating apparatus for 5 minutes at 1200 rpm. Due to the volatility of the BHT, the vials were heated for 30 minutes at 60°C to force the BHT into the headspace of the sealed vials. The headspace was exposed to a 65 μm DVB/PDMS SPME fiber punctured through the septum for 10 minutes. The SPME fiber bound the BHT present in the headspace during these 10 minutes. The fiber was then inserted into the inlet of a Torion Tridion-9 Portable GCMS to be heated and desorbed for analysis. Each spiked sample was prepared in triplicate. The calibration curves (with and without BHT) were created with the original unknown concentration of BHT in the spiked lotion matrix being $y = 0$.

3.2.4 Case study 1: head space needle trap method

The second method used a needle trap device (NTD) for binding the BHT. NTDs are needles with sorbent material on the inside of the needle that can bind volatile organic compounds (VOCs) or semi-volatile organic compounds (SVOCs) [14]. One end of the needle is connected to a suction pump and the other end of the needle remains open to draw gaseous sample into it to interact with the sorbent. The NTD can then be inserted into a GCMS inlet to be heated to desorb any bound chemical compounds.

The same calibration curve creation procedure with the SPME fiber in Section 3.2.3 was performed with the NTD with the setup portrayed in **Figure 3**. There were two needles inserted through the cap septum. The first needle was a purge needle

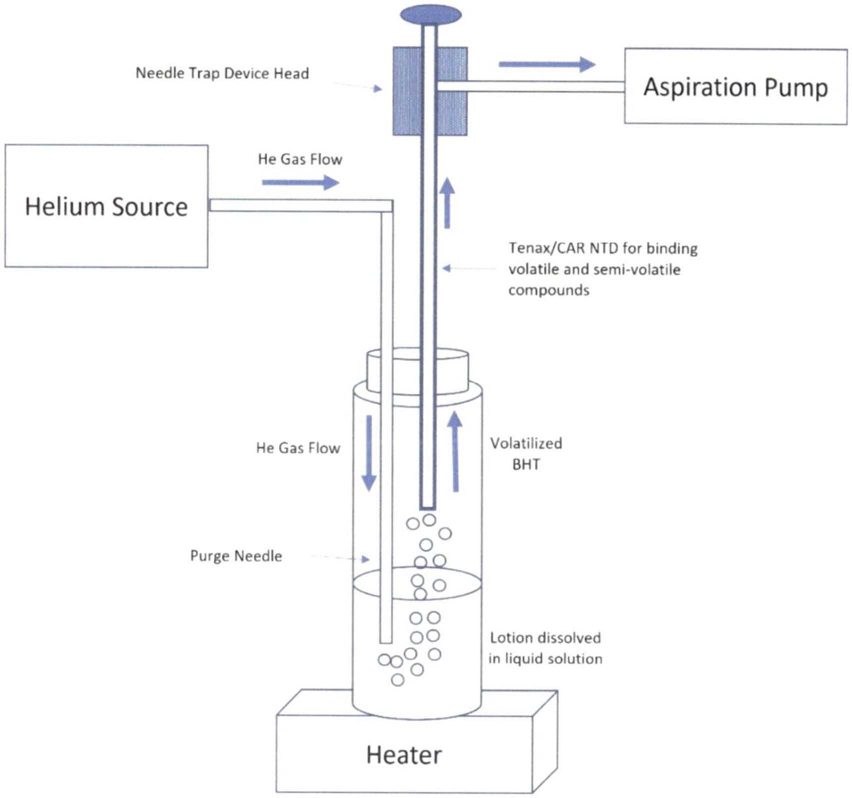

Figure 3.
NTD device setup.

which would flow helium through the spiked lotion solution. The BHT and other compounds would be volatilized easier with the gas flow through the solution. The second needle (NTD) was attached to a suction pump which would cause the head-space of the sample to be sucked through the sorbent. The NTD contained Tenax/CAR sorbent. Approximately 10 mL of gas passed through the NTD at a rate of 10 mL per minute. No breakthrough of the BHT occurred over this time and flow rate was based on the study coupling two NTDs in series previously. At one minute the run-time for this method was significantly less than the 10 minutes required for the SPME fiber to bind the BHT. The NTD was injected into a Torion-9 portable GCMS system. Again, the calibration curves created quantitated the original unknown BHT concentration in the lotion.

3.2.5 Case study 1: thin film liquid injection

Thin film (TF) liquid injection was used to help confirm the amounts of BHT extracted from the lotion in the standard addition curves. The TF method put small amounts of liquid with predetermined concentrations of analytes on a thin film membrane [15]. The membrane containing the liquid with the analyte of interest was placed into a SPS-3 PerkinElmer thermal desorption unit (TDU). The TDU device when connected to a NTD via a sorbent tube, transferred the analyte from the TF to the NTD. The NTD was then put into the portable GCMS inlet and heated at 250°C to desorb the BHT and analyze it in the portable GCMS. This allowed neat, liquid BHT standards to be injected into the instrument to correlate amount of BHT to the instrument response. A calibration curve was obtained when different BHT concentrations were injected using the TF method. BHT amounts put onto the TF ranged from 0.15 μg to 3 μg. This was done to confirm the amount of BHT extracted from the spiking with the standard curve creation and to observe any instrument response differences caused by interfering peaks from the lotion. This method also served as an external calibration method.

3.2.6 Case Study 1: results

The results for head space needle trap sampling coupled to a portable GCMS for testing BHT in cosmetics are shown in **Table 1**. The instrument responses from the TF BHT method were used to transform the NTD and SPME extraction responses to the amount of BHT injected on column versus spiked BHT standard extracted from the lotion. These values allowed for the quantitation of the BHT in the lotion using headspace/SPME BHT. The retention time of the BHT in the GCMS method was 95.4 seconds with no co-eluting peaks detected when comparing the standard addition curves of lotion with and without BHT.

Injecting three replicates showed that variability was less than 10% and had good linearity with a R^2 value of 0.98. As mentioned above, the NTD and SPME standard addition curves that originally had response versus spiked BHT standard were transformed to the amount of BHT extracted versus spiked BHT standard. The amount of BHT originally in the lotion was calculated when the value of y was set to zero in $y = mx + b$. There was good agreement between the NTD and SPME standard curves with only a 7.4% difference between the slopes of each curve. Spiked recoveries and comparison between the NTD standard addition and the external calibration are shown in **Table 1**.

A 0.005% BHT concentration was estimated based on the data from the standard addition curves. The results show that the NTD was able to concentrate the BHT from the headspace effectively even when mixed with the lotion matrix. Even with dilution by the helium, the headspace NTC method was able to achieve less than 10% relative standard deviation between replicates. Extracting from the headspace was a method

	BHT (μg)	**BHT (μg)**	
Non-BHT containing body lotion	Spiked	Recovered	Extraction Efficiency
Sample 1	7	6.5	92.8
Sample 2	5	5.4	108.0
Sample 3	15	15.2	101.3
	Calculated from standard addition curve	**Calculated from external calibration curve**	
BHT containing body lotion	5.6	6.1	108.9

Table 1.
Spike recoveries and comparison of NTD standard addition curve and external calibration.

advantage because some protocols may call for a dilute and shoot GC method for the lotion which would be problematic for the instrumentation.

3.2.7 Case 1 study: conclusion

The new method described in this case study was a headspace needle trap method used for the rapid determination of BHT using a portable GCMS. The sample preparation used a purge and trap method that included only diluting the lotion sample in solvent and then enriching the BHT sample on the NTD sorbent. Laborious sample preparation procedures, such as liquid-liquid partitioning, were avoided to obtain a simpler and faster procedure. The NTD method results agreed well with the SPME results which further strengthened the method validity. In addition, due to the air flow across the NTD, theoretically the NTD can concentrate more analyte in a shorter amount of time compared to SPME.

Future work suggested in this study would be a longer extraction NTD time with higher temperatures to make the method more exhaustive if the sample matrix is changed. This method allowed for rapid determination of BHT in non-laboratory environments lending to better quality control in factories or storage facilities if inspections of those locations were required.

3.3 Case study 2

3.3.1 Case study 2: introduction

Adulterants may be added to illicit drugs to either boost the drug's effects or increase profits [16]. The mixing of different compounds together can lead to unforeseen negative side effects if ingested, even leading to death. In addition, these adulterants can also be mixed with ingredients that do not have any pharmacological effect such as sugars and bicarbonates [16].

Identifying adulterants in crime labs is often not done due to the lack of analytical method development for these compounds and the adulterants are not considered illegal compounds of abuse [16]. Including these compounds in the analytical methods would benefit clinicians when diagnosing the treatment for acute toxicity from drug mixtures, help investigate chronic health impacts of these mixtures and even serve as chemical fingerprints when tracking illicit drugs since certain illicit drug manufacturers have signature recipes [16].

This study was meant to develop and validate a method for illicit drug analysis along with potential adulterants using a FLIR Griffin G510 portable GCMS system and then compare the results to a laboratory based GCMS system.

3.3.2 Case study 2: methodology

A FLIR Griffin G510 portable GCMS was used in this study. A 5 m × 0.18 mm × 0.18 μm DB-5 column was the low thermal mass GC column used for analytical separation. The carrier gas was helium supplied by the internal helium cartridges of the G510. The programmed temperature gradient was as follows: ramp of 30°C per minute from 50°C to 340°C and hold at 340°C for 4 min (total runtime of 13.6 minutes). Full scan mode was used (43 to 425 m/z) with 275°C injection port temperature and a 1 μL splitless injection.

An Agilent GCMS 6890 N/5975B was used as the laboratory based GCMS to confirm the test results from the portable GCMS system. This GCMS was operated in full scan mode from 40 to 550 m/z. The GC column was a DB-1 column (12 m X 22 mm X 0.3 µm). The following parameters were used: 1 µL injection volume, splitless mode injection at 265°C and detector at 300°C with a 1.2 mL/min helium flow rate. The temperature program for the GC method was as follows: ramp 30°C per minute from 50°C to 340°C and hold at 340°C for 2.33 min (total runtime of 12 minutes).

The main compounds of interest were as follows: alprazolam, amphetamine, aminopyrine, benzocaine, caffeine, cocaine, codeine, diltiazem, ephedrine, fentanyl, fenethylline, furanylfentanyl, heroin, hydroxyzine, levamisole, lidocaine, methamphetamine, morphine, noramidopyrine (a marker of metamizole), phencyclidine, phenacetin, procaine, strychnine and xylazine. Stocks were made to a concentration of 1 mg per mL and diluted to 0.1 mg per mL for method validation. Each solution was injected separately to establish retention time and confirm NIST library match scores. To mimic field conditions, a protocol was made to weigh approximately 1 mg of powder into 10 mL of methanol to give a 0.1 mg per mL concentration. This was an approximate concentration since analytical balances would not be brought into the field. The five main parameters to create a validated method for the 24 above compounds on a portable GCMS system were interference, precision, limit of detection, robustness, and carryover.

To analyze precision, each of the target compounds was injected 10 times every day for 3 days. In addition, fresh solutions were made every day. Confirmation of each peak included setting a 0.3-minute retention time variation limit and a NIST library match score of at least 65 or above. If the criteria were not met, then the peak seen in the GC run was not identified as one of the 24 compounds of interest.

Possible interferences would be the adulterants, and according to this study they were commonly found in seized illegal drugs [16]. The interferences evaluated were as follows: salicylic acid, atropine, cannabidiol, delta 9-THC, diphenhydramine, ibuprofen, methadone, mitragynine, nicotine, quinine, lactose, creatine, acetaminophen, thebaine and theophylline [16]. Stock solutions of 0.1 mg per mL were made and analyzed in duplicate to establish retention times and the most abundant mass fragments from the mass spectrum. The mass spectrum allowed for the identification of any unique fragments for each compound which showed selectivity in differentiating between signal features.

Limit of detection (LOD) was established by analyzing the lower concentrations to identify the point when the signal to noise ratios were less than 3:1. This reproducible 3:1 instrument response plus the 0.3-minute retention time window and the NIST score greater than or equal to 65 were all criteria in identifying the LOD and confirming the identity of the compound. All these values were used to help with the validation of this method.

The carryover was tested by injecting 0.2, 0.5, and 1.0 mg per mL of the drugs of abuse and the adulterants. Two solvent blanks (methanol) were then injected in duplicate. The concentration at which the method did not have carryover was designated as the highest analyte concentration at which none of the illicit drug samples or adulterants were observed in the blank.

To evaluate the robustness of the instrument or the portable GCMS instrument performance reliability, small changes were made to see any fluctuations in instrument response. The injector temperature was varied plus or minus 5°C and the injection volume was varied plus or minus 0.2 µL. Each of the variations was done in duplicate and the criteria for peak identification of each of the 24 analytes along with the adulterants were retention time windows of 0.3 minutes and GCMS library match scores greater than or equal to 65.

After validation of the portable GCMS method, tests were performed to show the method could correctly identify the drugs of abuse and adulterants. Fifty different seized drug lots were tested for this purpose. The benchtop GCMS results were run alongside the portable GCMS to confirm the reliability of drug testing results from the portable GCMS. This study described using a method called Receiver Operating Characteristic (ROC) analysis to show the reliability of the portable GCMS results [16]. The ROC analysis assigned different categories to the portable GCMS data when it was compared to the benchtop GCMS. As above, identification criteria were established with specific parameters related to retention time and GCMS library identification score. True positive (TP) samples showed positive identifications on both the portable GCMS and benchtop GCMS for target analytes. True negative (TN) samples showed negative identifications on both portable and benchtop GCMS. False positive (FP) samples screened positive on the portable GCMS for a target analyte, but it was confirmed absent on the benchtop GCMS. False negative (FN) results showed an absence of target analytes on the portable GCMS results, but positive identification of the targets was confirmed on the benchtop GCMS. Based on the results, each compound in the 50 seized drug lots received a TP, TN, FP, or FN designation, and these values were used to calculate the following performance parameters describes in **Table 2**: sensitivity, specificity, accuracy, positive predictive values (PPV), and negative predictive values (NPV).

3.3.3 Case study 2: results

Some of the 24 target compounds had similar retention times to several adulterants, but the mass spectrum of each of the co-eluting peaks were distinct enough to resolve/differentiate these compounds. There was no carryover of the analytes of interest or adulterants using the concentrations described in the previous section. Carryover was important to test since sample concentration variability was expected to be high in the field without calibrated glassware or analytical balances. Confirming the absence of carryover ensured a lower chance of false positives when there was high sample preparation variability causing the concentration to be greater than 1 mg/mL. Limits of detection ranged from 0.01 to 0.1 mg/mL depending on the analytes of interest in this study.

Parameter	Formula (%)	Definition
Sensitivity	100 × (TP#) / (TP# + FN#)	% Positively identified results in confirmed positive samples
Specificity	100 × (TN#) / (TN# + FP#)	% Negatively identified results in confirmed negative samples
Accuracy	100 × (TP# + TN#) / (TP# + TN# + FP# + FN#)	% Similarity between portable GCMS and benchtop GCMS
Positive Predictive Value (PPV)	(TP#) / (TP# + FP#)	Ratio of true positive results
Negative Predictive Value (NPV)	(TN#) / (FN# + TN#)	Ratio of true negative results

Table 2.
Performance parameters use to score portable GCMS performance.

Thirty injections at 0.1 mg/mL of each of the standards were made to measure the precision. Heroin and morphine were not detected in 1 out of the 30 injections while diltiazem and fenethylline were not detected in 3 out of 30 injections. The precision for the other analytes were not mentioned. Many aspects of manual injection can change the results such as air bubbles, injection speed, and injection timing. It is important for the analyst in the field to emphasize reproducible injection technique.

Fifty seized illegal drug samples composing of mainly cocaine, methamphetamine, and heroin underwent screening with the validated portable GCMS method using the FLIR Griffin 510. Data was confirmed with a benchtop GCMS system. The data showing the results of the Receiver Operating Characteristic (ROC) are in **Table 3**. The data confirmed the presence or absence of adulterants and other illicit drugs mixed with the cocaine, methamphetamine, and heroin samples. Only 5 samples (10%) were true positives for one substance and 19 samples (38%) had two substances from the above list. Ten samples (20%) contained three of the above compounds, and 16 samples (32%) had four or more of the above drugs and or adulterants. The study did not specify which samples contained the adulterants. The data did not include all the compounds developed for the portable GCMS method because the samples tested in **Table 3** only contained the compounds with true positive or false negative hits. Refer to the methods for definitions of TP, FN, FP, and TN.

Some additional facts of the data need to be discussed. Accuracy values from the portable GCMS that were in high agreement with the lab based GCMS had higher values in the accuracy column. As Noted in from the methods section, the sensitivity, accuracy, and NPV were affected by the number of false negatives as seen with caffeine, heroin, procaine, amphetamine, lidocaine, and benzocaine. Most compounds had high values, but each method had limitations such as benzocaine having a PPV that could not be calculated since there were no true positive samples.

3.3.4 Case study 2: conclusion

The screening method developed in this study measured 24 different illicit drugs along with several adulterants on a FLIR Griffin 510 portable GCMS. To confirm the results, a laboratory based GCMS was used to set criteria for measuring sensitivity, specificity, accuracy, and the probability of predicting whether a compound would be present or absent. Through this process, the portable GCMS method was validated to show a relatively high degree of accuracy for correctly screening the presence or absence of these compounds in 50 seized drug samples. The adulterants did not significantly affect the performance of the method, and this method was prepared for monitoring of these adulterants as needed.

3.4 Case study 3

3.4.1 Case study 3: introduction

The need exists to detect and quantitate chemical warfare agents (CWAs) in the field in order to protect those who may be at risk of exposure. The portable GCMS systems are most advantageous since deployment in the field yields shorter data turnaround times as samples do not have to be transported to laboratories for analysis [17, 18]. Besides short turnaround time, reproducible CWA absolute

Drug	TP#	FN#	FP#	TN#	Sensitivity	Specificity	Accuracy	PPV	NPV
Amphetamine	7	1	0	42	87.5	100.0	98.0	100.0	97.6
Benzocaine	0	1	0	49	—	100.0	98.0	—	98.0
Caffeine	13	4	0	33	76.4	100.0	92.0	100.0	89.1
Cocaine	20	0	0	30	100.0	100.0	100.0	100.0	100.0
Diltiazem	4	0	0	46	100.0	100.0	100.0	100.0	100.0
Ephedrine	2	0	0	48	100.0	100.0	100.0	100.0	100.0
Fentanyl	2	0	0	48	100.0	100.0	100.0	100.0	100.0
Furanyl Fentanyl	3	0	0	47	100.0	100.0	100.0	100.0	100.0
Heroin	12	3	0	35	80.0	100.0	94.0	100.0	92.1
Hydroxyzine	3	0	0	47	100.0	100.0	100.0	100.0	100.0
Levamisole	7	0	0	43	100.0	100.0	100.0	100.0	100.0
Lidocaine	2	1	0	47	66.6	100.0	98.0	100.0	97.9
Methamphetamine	15	0	0	35	100.0	100.0	100.0	100.0	100.0
Morphine	2	0	0	48	100.0	100.0	100.0	100.0	100.0
Noramidopyrine	3	0	0	47	100.0	100.0	100.0	100.0	100.0
Phenacetin	9	0	0	41	100.0	100.0	100.0	100.0	100.0
Procaine	2	2	0	46	50.0	100.0	96.0	100.0	95.8

Table 3.
Results comparing Portable GCMS data to benchtop GCMS results. Sensitivity, specificity, accuracy, PPV, and NPV were % values.

quantitation data for portable GCMS is very important due to the high toxicity of these substances. Shown in **Figure 4**, the analytes of interest in this study were the G-series nerve agents tabun, sarin, soman, and cyclosarin and the blistering agent sulfur mustard [18]. Previous work showed the possibility of using response factors (analyte peak area/internal standard peak area) for quantitating G-series nerve agents using Hapsite and Hapsite ER portable GCMS systems [19]. The results showed significant carryover effects from concentrator sorbent, air sampling robe, and transfer line. In addition, the Hapsite ER system built-in internal standard, bromopentafluorobenzene (BPFB) showed 26.3% relative standard deviation (%RSD) between days and 32.9% RSD within a day. Both high % RSD values showed that the BPFB was not a good candidate for calculating relative response factor CWA calibration curves.

This case study sought to show that better? %RSD values for the above compounds could be achieved by spiking potential candidate compounds (**Figure 5**) onto thermal desorption (TD) tubes. These compounds served as focusing agents allowing relative response factors (RRFs) to be used to create calibration curves with better %RSD values and thus better quantitation parameters for portable GCMS systems [18]. The focusing agents in **Figure 5** were as follows: 2-chloroethyl ethyl sulfide (2-CEES), diisopropyl fluorophosphate (DIFP), diethyl methylphosphonate (DEMP), diethyl malonate (DEM), methyl salicylate (MES), and dichlorvos (DCV). In addition, the stability of the focusing agents 14 days after spiking was also determined at multiple conditions. Another goal was to reproduce the data across multiple Hapsite portable GCMS systems to show inter-instrument feasibility for low % RSD quantitation of the CWAs.

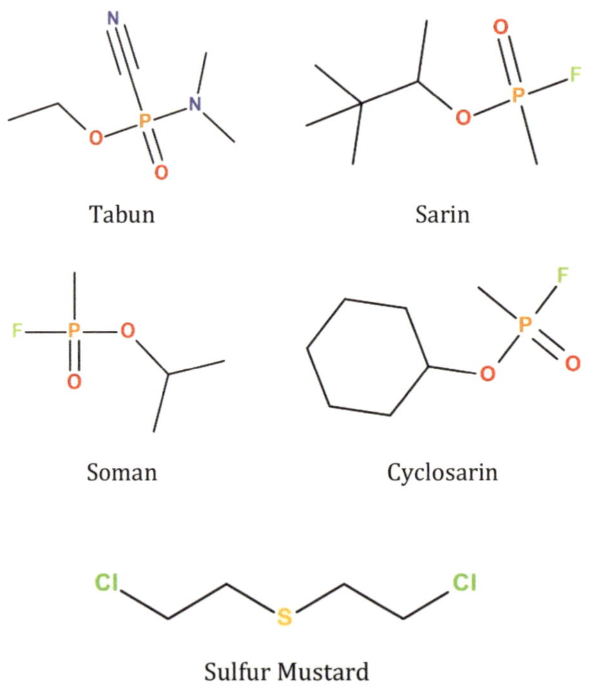

Figure 4.
Nerve agents and blistering agents.

3.4.2 Case study 3: methodology

The study used Supelco Tenax® TA (35/60) TD tubes to measure sarin, tabun, soman, cyclosarin, and sulfur mustard on multiple Hapsite ER portable GCMS systems. The TDU sampling system was set to 310°C and nitrogen gas was used to transfer the desorbed sample to a tri-bed concentrator. This concentrator was then held at 45°C for 12 minutes, 280°C for 11 seconds, and then the desorbed sampled entered a DB-1 ms GC column (15 m, 0.25 mm ID, 1.0 μm d_f). The total run time of the method was 15 minutes 30 seconds. The programmed temperature method was as follows: hold at 60°C for 1 minute 15 seconds, ramp of 8°C per minute for 3 minutes 45 seconds, after reaching 90°C ramp at 25°C per minute for 4 minutes 24 seconds with a maximum temperature at 200°C. Hapsite portable GCMS system were used with scanning from 45 to 300 m/z [18].

The TD tubes were spiked with 1, 2, 5, 10, and 50 ng of each of the previously mentioned focusing agents. Stability, carry-over, and remaining residual focusing agent were tested with two Hapsite ER portable GCMS systems. Carryover was determined by desorbing a spiked TD tube and then desorbing a blank TD tube. Residual focusing agent remaining on the TD tube was analyzed by desorbing spiked TD tubes multiple times. Stability was determined by measuring the mid-point signal response of 5 ng of each of the focusing agents spiked on a TD tube over a 14-day time period with multiple time points in between. External calibration curves were made with the system internal standard but had a %RSD value of around 24.2%. RRFs used area values determined by Hapsite ER IQ software and Automated Mass Spectral

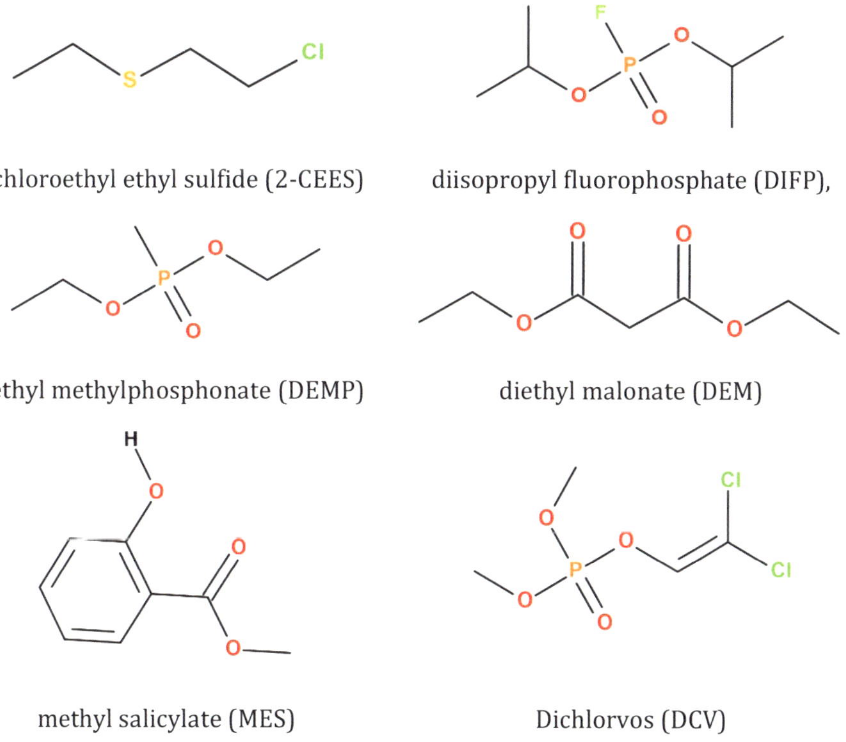

Figure 5.
Structures of focusing agents used in study.

Deconvolution and Identification System (AMDIS) version 2.72 [18]. TD tubes were conditioned before spiking with dry purging rates of 50 mL of nitrogen per minute and temperature hold of 280°C for 120 minutes.

RRFs were calculated by dividing the (CWA Area X mass of focusing agent) by the (Focusing agent area X mass of the CWA). Six-point calibration curves were made for 1, 2, 5, 10, 20, and 50 ng of each CWA in relation to the RRF of each CWA and focusing agent.

3.4.3 Case study 3: results

The goal of this study was to improve the quantitation method of sarin, tabun, soman, cyclosarin, and sulfur mustard on a portable GCMS using the focusing agents listed above. The retention times of the CWAs and the focusing agents were mostly chromatographically resolved. Soman (retention time 6 minutes 13 seconds) was close to DEMP (retention time 5 minutes 52 seconds) and DEM (retention time 6 minutes 31 seconds). Cyclosarin (retention time 7 minutes 44 seconds) was not chromatographically resolved from focusing agent MES (retention time 7 minutes 48 seconds) in the total ion chromatogram, but the quantitation was done on specific and distinct mass spectral features [18]. If the mass spectral data was not present, quantitation of cyclosarin would be potentially less accurate.

All CWAs and focusing agents were analyzed on a single TD tube with 5 ng of each focusing agent. The measured peak areas were used to calculate the RRFs and calibration curves for each of the CWAs in triplicate across four Hapsite ER portable GCMS systems. Carry over of the CWAs was established to be low. For tabun, sarin, soman, and cyclosarin, the carryover was as follows: 0.26%, 0.04%, 0.01% 0.02%. Carryover for sulfur mustard was not given in the study. The residual BPFB internal standard was around 58%, and the carryover was less than 0.1%.

All RRFs were combined for six-point calibration curves for each of the CWAs of 1, 2, 5, 10, 20 and 50 ng of each CWA. The RRFs for the calibration curves were tested over 14 days and the % RSD values of each of the CWAs with each of the focusing agents are included in **Table 4**. All calibration curves were linear and had R^2 values of at least 0.983.

All % RSD values were compared to the % RSD of the internal BPFB internal standard of 32.9%. Most of the %RSD values were lower than the 32.9% except for the CWAs with 2-CEES. Sulfur mustard had high % RSD for most of the focusing agents but MES seemed to be the best for this specific compound. This was a significant improvement over the BPFB and thus allowed for greater confidence in the CWA

	%RSD					
CWA	**2-CEES**	**DIFP**	**DEMP**	**DEM**	**MES**	**DCV**
Tabun	45	14	12	22	30	20
Sarin	51	3	8	10	21	9
Soman	37	18	25	9	2	9
Cyclosarin	41	12	18	2	9	3
Sulfur Mustard	29	32	37	22	12	23

Table 4.
Average RRF %RSD values for CWAs over 14 days.

quantitation. DIFP, DEMP, and DEM had the lowest overall %RSD for most of the CWAs. The overall difference in area responses for all CWAs between two Hapsite instruments were as follows: 2-CEES 21%, DIFP 6%, DEMP 6%, DEM 16%, MES 13%, and DCV 7%. The similar values for DIFP and DEMP showed that the calibration curves were transferable between Hapsite instruments and were not just limited to one portable GCMS.

3.4.4 Case study 3: conclusion

This study showed that thermal desorption calibration techniques on the Hapsite portable GCMS system could be used to improve quantitation reproducibility of CWAs. By using focusing agents, the intra-instrument relative standard deviation was greatly reduced below the 32% using the internal BPFB standard. In addition, the inter-instrument variability was also shown to be much lower than that of the BPFB standard. This study successfully showed that CWA quantitation on portable GCMS equipment in the field can have high reproducibility and thus be more valid as absolute quantitation numbers.

4. Conclusions

As shown in this chapter, many different fields such as manufacturing, government, and first responders elect to use portable gas chromatography mass spectrometers (GCMS) for analysis. For each field, compound analysis targets are different, but testing is required for public safety such as with illegal drugs, chemical weapons, or potentially harmful ingredients in cosmetics. These systems are used for their mobility and sensitivity for analyzing complex mixtures in the field. Without portable GCMS technology, toxic samples may have to be shipped to laboratories which would delay data analysis and decision making.

Section 2 of this chapter reviewed the technology behind portable GCMS systems. The GC columns are low thermal mass column technology allowing for miniaturization and 1% the power requirements of a laboratory based GCMS. The following commercial portable GCMS systems were discussed in the chapter: Perkin Elmer Torion T-9series, FLIR Griffin 500 series, and Infincon Hapsite ER series. All have similar specifications, but slightly different sample introduction and sample preparation technologies. The characteristics of each system require user evaluation to choose which system would be better for their specific methods and deployment criteria.

Section 3.1 reviewed a Homeland Security report on the portable GCMS characteristics which were important to users in the first responder fields. Many different evaluation tests were done to evaluate the three different portable GCMS systems. Many of the criteria were similar on the different systems, but ease of use and deployment were important to most users. Again, a user evaluation would be required before purchase to match system requirements to user method development requirements.

Section 3.2 contained a method for the quantitation of butylated hydroxy toluene using a portable GCMS and needle trap technology for a purge and trap method. This work created a method which would allow researchers to decrease the time it would take for quality control testing.

Section 3.3 described the development of a new method for evaluation of 24 different illicit drugs with adulterants which could interfere with quantitation using portable GCMS technology. Method development showed that the portable GCMS results were the same as the benchtop GCMS technology in a laboratory.

Section 3.4 reviewed that focusing agents used on TDU tubes allowed for better precision when measuring the chemical weapon agents on portable GCMS systems. Reliable quantitation of these compounds was vital when the lives of personal would be put in danger. Also, depending on the application needs, these instrument can perform in-field analysis just as well as instruments in laboratory settings.

Conflict of interest

The authors declare no conflict of interest.

Author details

Robert Owen Bussey III
RJ Reynolds Tobacco, Winston-Salem, NC, USA

*Address all correspondence to: busseyr@rjrt.com

References

[1] Field Portable Gas Chromatograph Mass Spectrometers Assessment Report [Internet]. 2020. https://www.dhs.gov/sites/default/files/saver-gcms_assessment_report_05feb2020-508.pdf. [Accessed: July 23, 2022]

[2] McNair H, Miller J, Snow N. Basic Gas Chromatography. 3rd ed. New York: John Wiley and Sons, Inc; 2019. DOI: 10.1002/9781119450795

[3] Luong J, Gras R, Mustacich R, Cortes H. Low thermal mass gas chromatography: Principles and applications. Journal of Chromatographic Science. 2006;**44**:253-261. DOI: 10.1093/chromsci/44.5.253

[4] Product Note: Torion T-9 GC MS [Internet]. 2016. Available from: https://resources.perkinelmer.com/lab-solutions/resources/docs/PRD_Torion_T-9_GCMS_(012311C_01).pdf?_ga=2.71701523.1052543098.1659799764-1374782003.1657242074. [Accessed: August 7, 2022]

[5] Torion T-9 GC MS Product Overview [Internet]. 2016. Available from: https://www.perkinelmer.com/product/torion-t-9-portable-gc-ms-instrument-ntsst090500. [Accessed: August 7, 2022]

[6] Novel Coiled Microextraction Sampling Device used for Field Sampling of Illicit Drugs of Abuse and Analysis by Micro Gas Chromatograph/Mass Spectrometer. Chromatography Today. [Internet]. 2021. Available from: https://www.chromatographytoday.com/article/sample-prep/67/perkinelmer/novel-coiled-microextraction-sampling-device-used-for-field-sampling-of-illicit-drugs-of-abuse-and-analysis-by-micro-gas-chromatographmass-spectrometer/3007. [Accessed: August 7, 2022]

[7] GRIFFIN™ G510 Person-Portable GC-MS Chemical Brochure [Internet]. 2021. Available from: https://flir.netx.net/file/asset/8333/original. [Accessed: August 7, 2022]

[8] FLIR GRIFFIN™G500 SERIES [Internet]. 2021. Available from: https://flir.netx.net/file/asset/35809/original. [Accessed: August 7, 2022]

[9] HAPSITE® ER Chemical Identification System and Accessories [Internet]. 2022. Available from: https://www.inficon.com/en/products/hapsite-er-identification-system. [Accessed: August 7, 2022]

[10] Ghosh C, Singh V, Grandy J, Pawliszyn J. Development and validation of a headspace needle-trap method for rapid quantitative estimation of butylated hydroxytoluene from cosmetics by hand-portable GC-MS. RSC Advances. 2020;**10**:6671-6677. DOI: 10.1039/C9RA08676E

[11] Bogdan Allemann I, Baumann L. Antioxidants used in skin care formulations. Skin Therapy Letters. 2008;**13**:5-9

[12] Bronaugh RL, Collier SW, Storm JE, Stewart RF. In vitro evaluation of skin absorption and metabolism. Journal of Toxicology: Cutaneous and Ocular Toxicology. 1989;**8**:453-467. DOI: 10.3109/15569528909062950

[13] Fries E, Püttmann W. Analysis of the antioxidant butylated hydroxytoluene (BHT) in water by means of solid phase extraction combined with GC/MS. Water Research. 2002;**36**:2319-2327. DOI: 10.1016/S0043-1354(01)00453-5

[14] Wang A, Fang F, Pawliszyn I. Sampling and determination of volatile

organic compounds with needle trap devices. Journal of Chromatography A. 2005;**1072**:127-135. DOI: 10.1016/j.chroma.2004.12.064

[15] Grandy J, Singh V, Lashgari M, Gauthier M, Pawlliszyn J. Development of a hydrophilic lipophilic balanced thin film solid phase microextraction device for balanced determination of volatile organic compounds. Analytical Chemistry. 2018;**90**:14072-14080. DOI: 10.1021/acs.analchem.8b04544

[16] Fiorentin TR, Logan BK, Martin DM, Browne T, Rieders EF. Assessment of a portable quadrupole-based gas chromatography mass spectrometry for seized drug analysis. Forensic Science International. 2020;**313**:110342. DOI: 10.1016/j.forsciint.2020.110342

[17] Leary PE, Kammrah BW, Lattman KJ, Beais GL. Deploying Portable gas chromatography-mass spectrometry (GC-MS) to military users for the identification of toxic Chemical agents in theater. Applied Spectroscopy. 2019;**73**:841-858. DOI: 10.1177/0003702819849499

[18] Kelly JT, Qualley A, Hughes GT, Rubenstein MH, Malloy TA, Piatkowski T. Improving quantification of tabun, sarin, soman, cyclosarin, and sulfur mustard by focusing agents: A field portable gas chromatography-mass spectrometry study. Journal of Chromatography A. 2021;**1636**:461784. DOI: 10.1016/j.chroma.2020.461784

[19] Nagashima H, Kondo T, Nagoya T, Ikeda T, Kurimata N, Unoke S, et al. Identification of chemical warfare agents from vapor samples using a field-portable capillary gas chromatography/membrane-interfaced electron ionization quadrupole mass spectrometry instrument with tri-bed concentrator. Journal of Chromatography A. 2015;**1406**:279-290. DOI: 10.1016/j.chroma.2015.06.011

Chapter 4

Liquid Extraction for Flavor and Fragrance Analyses in Consumer Products

Zhigang Hao, Vivian Liu, Jake Salerno, Yu Wang, Mania Bankova and Long Pan

Abstract

Gas chromatography-mass spectrometry is a powerful tool to analyze flavor and fragrance from raw materials to the final commercial products. During the development of new technologies, most focuses have been given to novel columns, advanced detectors, and automation designs to leverage the instrument capabilities. The fundamental factors including polarity impact on sample homogenization and chemical interaction between analytes and extraction solvents are not equally emphasized during the sampling procedures. The current project focused on the liquid extraction procedures prior to GCMS analysis. Significant nucleophilic reactions were found to take place when a water-ethanol solvent was tried to extract flavor and fragrance ingredients. The isooctane in water-isooctane extraction system is friendly with GC columns and effective to extract the volatiles. However, the surfactants, humectants, and polymers in consumer cleaning products have significant impact on analyte distribution between water and isooctane solvents. The enhanced solubility of certain ingredients in water phase will change their profiling information in isooctane. During such extractions, hydrophilic volatile ingredients can be missed and the results become unreliable. For this reason, a newly designed water-n-propanol-isooctane extraction system was compared. This one-phase sample solution follows the homogenization rule in analytical chemistry and be more representative to the original samples.

Keywords: liquid extraction, gas chromatography, mass spectrometry, flavor, fragrance, volatiles, consumer products

1. Introduction

Personal and oral care products that serve subtle human needs are tied inextricably to sensory values of taste, odor, and texture. Moreover, failure to meet flavor and fragrance expectations may often signal poor quality and could even be related to inherently subjective influence. For these commercial reasons, product manufacturers, distributors, buyers, wholesale and retail sellers, and especially consumers need reliable ways to assess product flavor and fragrance quality. From the perspective of a consumer product company, "a reliable assessment" of a chemical mixture—whether it is raw materials or final products—calls for the need to make a correct analytical measurement. A "reliable assessment" should be objective; even the term "flavor and fragrance" could often be inherently subjective [1].

Flavor and fragrance are typically consistent with volatile ingredients. The determination of volatile components in a mixture is a process widely used in many disciplines, such as flavor, fragrance, environmental, food, forensic, oil, pharmaceutical, and consumer product analysis [2, 3]. The method of choice for many of these analyses can be simply described as a sampling procedure plus instrumental analysis such as GCMS. Solvent extraction is the most common sampling step before the GCMS analysis is performed. During extraction, all volatile ingredients should be transferred and dissolved into extraction solvents. Qualitative and quantitative studies of chemical compounds from different matrices, such as plant materials, drugs, and consumer products, rely mostly on the selection of appropriate extraction methods [4, 5]. Extraction plays a significant and crucial role for the final outcome. Extraction methods are sometimes referred to as "sample preparation techniques." Most of the time, this part of the study is neglected and done by non-trained research personnel [6, 7], despite two-thirds of the analytical chemist workforce accounting for sample preparation techniques. Most researchers believe in the importance of sample preparation during any analytical studies because the analytes could be missed or alternated without a suitable sampling procedure [8]. Even with very selective mass detectors such as Thermo orbitrap or time of flight, sample cleanup procedure is still a critical stage for the final analytical results [9]. It is true that the development of modern chromatographic and spectrometric techniques makes chemical compound analysis easier than before, but a successful method still heavily depends on sampling procedures, input parameters, and exact nature of materials [10]. High-resolution mass detectors can filter out a large part of sample matrix noise due to its mass selectivity and ensure more precise results. However, the detector cannot guarantee the high accuracy if certain ingredients are not fully extracted and transferred into the instrument *via* solvents during the extraction period.

Common factors affecting extraction processes are sample matrix property, solvent selection, operation temperature, pressure, and time [11]. Those common factors could cause the chemical interaction between extraction solvents and chemical composition to be analyzed. The chemical interaction could not only cause chemical structure changes such as chemical reaction but also alternate a substance solubility, precipitation, solvation, complexation, and dissociation [12]. For cleaning consumer products such as toothpaste, the matrix usually contains a significant amount of silica base, humectants such as glycerin or sorbitol and polymers such as polyethylene glycol (PEG) and xanthan gum, and certain amount of surfactants. Silica usually has poor solvent solubility, and polymers usually contain both hydrophobic and hydrophilic segments inside their structures. That is why these products cannot be evenly suspended with organic solvents or even very polar alcohols such as methanol or ethanol. Water is necessary to suspend the toothpaste matrix and provide a homogenized sampling procedure. For this reason, we often use common extraction solvents such as methanol or ethanol with water to directly extract analytes from oral and personal care products for GCMS analysis [13]. In addition, highly hydrophobic organic solvents such as isooctane (2,2,4-trimethylpentane) were used to extract flavors and fragrance from the water phase with a fractionation operation. Those two methods have been alternatively used for the quantification of flavor and fragrance based on their polarity difference. However, those two methods faced the challenges during flavor and fragrance profiling projects for quality evaluation. Some existing ingredients were missing in GCMS chromatograms due to two extraction phases and ingredient distribution impacted by the sample matrix. The chemical interactions between flavor and fragrance ingredients and extraction solvents can change flavor and fragrance profiles during their chemical analysis. To overcome those challenges, a water organic miscible solvent system was developed and used for flavor and fragrance extraction in cleaning products such as toothpaste. In this study, three extraction procedures were explored and compared. Their positive and negative attributes were discussed. A general extraction summary is provided at the end.

2. Materials and methods

2.1 Chemical reagents and materials

Isooctane (2, 2, 4-Trimethylpentane, C_8H_{18}) was purchased from VWR (Radnor, PA, USA). Anhydrous sodium sulfate (Na_2SO_4) was purchased from Sigma-Aldrich (MilliporeSigma, St Louis, MO, USA). n-propanol (n-propyl alcohol, C_3H_8O) was purchased from Honeywell (Charlotte, NC, USA). Wintergreen oil (CAS#: 90045–28-6), lemongrass oil (CAS#: 8007-02-1), and peppermint oil (CAS#: 8006-90-4) were purchased from MilliPoreSigma). PTFE membrane filter was obtained from VWR.

2.2 Instrumentation

A Genie 2 vortex mixer from MilliPoreSigma was used to assist sample dispersion in a minimal amount of time. The Eppendorf centrifuge 5810R from MilliPoreSigma was used to centrifuge the toothpaste sample for 10 min at 5000 g before the GCMS analysis.

A gas chromatography system 6890 N combined with 5975 mass spectrometry detector (MSD) from Agilent Technology (Santa Clara, CA, USA) plus multiple-Purpose-Sampler (MPS2) from Gerstel (Linthicum, MD, USA) was used for flavor analysis. Separation was accomplished using the GC column HP-5MS (30 m x 0.25 mm x 0.25 μm, length x inside diameter x film thickness, Agilent Technologies). A 1 μL sample was injected in the splitless mode. The oven temperature was initially held at 80°C for 1 min. Thereafter, the temperature was raised at 4°C/min until 150°C and held for 1.5 min. Total running time is 20 min. Helium was used as the carrier gas and delivered at 1 mL/min constant flow rate. The gas pressure and velocity were at 8.2 psi and 37 cm/sec, respectively. The injector temperature was set at 250°C, and the interface temperature between GC oven and MS detector chamber was 250°C. The MS detectors were tuned with the standard spectrum auto-tune, and the MS data for total ion chromatogram (TIC) were acquired in the full scan mode (m/z of 29–450 at a scan rate of 3 scan/sec using electron ionization (EI) with electron energy of 70 eV. The MS source and quat temperatures were 230°C and 150°C, respectively.

2.3 Sampling procedure with ethanol-water system

2.3.1 Standard preparation using ethanol-water solvents

About 100 mg of flavor oil was added into 49.9 grams of a 1:8 water-ethanol solution. Total weight of the solution was 50.0 grams. Three flavor stock solutions (2000 ppm) from wintergreen, lemongrass, and peppermint oils were used as reference standards in this project. The stock solution was then diluted 20-fold with the 1:8 water-ethanol solution, reaching target concentrations of 100 ppm. This standard solution was transferred into a 2-mL autosampler vials for GCMS analysis.

2.3.2 Sample preparation using ethanol-water solvents

About 0.5 gram of toothpaste sample was weighed into a 50-mL polypropylene conical tube. The weight of 3.2 gram of water was added into the tube and mixed with toothpaste by using a Genie 2 vortex mixer into slurry. Next, a weight of 21.3 gram of ethanol was added to the tube. The solution was then mixed thoroughly with a Genie 2 vortex mixer, and the mixture was centrifuged by using Eppendorf centrifuge 5810R for 10 min at 5000 g. The top layer of solution was filtrated with a 0.2 μm PTFE membrane and transferred into a 2-mL autosampler vial for GCMS analysis.

2.4 Sampling procedure using isooctane with saturated sodium sulfate water solution

2.4.1 Preparation of a saturated sodium sulfate solution

About 40.0 gram of anhydrous sodium sulfate salt was transferred into a 250-mL Erlenmeyer flask with 100 mL of distilled water and mixed thoroughly with a magnetic stirring bar. After excess sodium sulfate started to rest at the bottom of the container, the upper clear saturated solution was carefully transferred to another clean glass jar for use.

2.4.2 Standard preparation using isooctane solvent

About 100 mg of flavor oil was added into 49.9 gram of isooctane. The total weight of the solution was 50.0 gram. Three flavor stock solutions (2000 ppm) from wintergreen, lemongrass, and peppermint oils were used as reference standards. Then, the stock solution was diluted 20-fold with isooctane solvent, reaching target concentrations of 100 ppm. This standard solution was transferred into 2-mL autosampler vials for GCMS analysis.

2.4.3 Sample preparation using isooctane with saturated sodium sulfate water solvents

About 0.5 gram of toothpaste sample was weighed into a 50 mL polypropylene conical tube. In total, 24.5 gram of saturated sodium sulfate water solution was added into the tube and toothpaste was mixed and suspended with a Genie 2 vortex mixer into the slurry. About 12.5 gram such aqueous toothpaste solution was mixed with 12.5 gram of isooctane solvent. This combination was then mixed thoroughly with the Genie 2 vortex mixer. The mixture was centrifuged by using Eppendorf centrifuge 5810R for 10 min at 5000 g. The top clear solution of isooctane was transferred into a 2-mL autosampler vial for GCMS analysis.

2.5 Sampling procedure with water-n-propanol-isooctane (1:8.5:15) solvent system

2.5.1 Standard preparation using water-n-propanol-isooctane (1:8.5:15) as solvent

About 100 mg of flavor oil was added into 49.9 gram of water-n-propanol-isooctane solvents. Total weight of the solution was 50 gram. Three flavor stock solutions (2000 ppm) from wintergreen, lemongrass, and peppermint oils were used as reference standards. The stock solution was then diluted 20-fold with water-n-propanol-isooctane solvents, reaching target concentrations of 100 ppm. This standard solution was transferred into 2-mL autosampler vials for GCMS analysis.

2.5.2 Sample preparation using water-n-propanol-isooctane (1:8.5:15) as solvent

About 0.5 gram of toothpaste sample was weighed into a 50-mL polypropylene conical tube. About 1.0 gram of water was added into the tube and toothpaste was mixed and suspended with the Genie 2 vortex mixer into the slurry. Next, 8.5 gram of n-propanol solvent was added. The mixture was then mixed thoroughly with the Genie 2 vortex mixer again. Lastly, 15.0 gram of isooctane was added. The combination was then blended thoroughly with Genie 2 vortex, and the mixture was centrifuged by using the Eppendorf centrifuge 5810R for 10 min at 5000 g. The top solution was filtrated with a 0.2 μm PTFE membrane and transferred into a 2-mL autosampler vial for GCMS analysis.

3. Results and discussion

3.1 Sampling procedure with ethanol-water system

Due to the specific properties of cleaning products we described in the introduction section, aqueous methanol and ethanol are very common liquid extraction solvent systems used for toothpaste sampling procedures [13, 14]. Water can easily suspend toothpaste into the slurry and organic solvents can effectively extract most analytes. The ratio between organic solvent and water can further determine the polarity of the analyte to be extracted. Because most flavor ingredients are hydrophobic, we used a high ratio of organic solvent at 1:8 ratio of water:ethanol to extract toothpaste flavors. The extraction efficiency and recovery were close to 100% as predicted because ethanol not only has good solubility for most flavor ingredients but also possesses penetration capabilities into toothpaste matrix materials such as silica and polymers. However, the hydroxyl group in ethanol structure has a very strong nucleophilic attaching capability, which can easily react with the compounds having carboxyl groups including esters. When quantification of methyl salicylate in wintergreen flavor and related products was tried with this extraction system, ethyl salicylate was identified within 30 min after sampling shown in **Figure 1**. From 30 min, 6 hours to overnight time periods, methyl salicylate continuously decreased and ethyl salicylate correspondingly increased, which is demonstrated well in **Figure 1**. In addition, when quantification of lemongrass flavor oil was tried with this same procedure, a new citral diethyl acetal peak at retention time of 15 min was observed and identified 60 min after sampling shown in **Figure 2**. Those reacted products indicated that the quantity of original flavor ingredients can be changed due to the reaction with extraction solvents. The final results on the GCMS instruments then become unreliable. Specifically, if an integrated flavor profile is needed to evaluate the flavor quality, the aqueous methanol and ethanol systems could become challenging.

3.2 Sampling procedure using isooctane with sodium sulfate saturated aqueous solution fractionation

Most volatile ingredients are hydrophobic or with a middle range of polarity. Usually, they can be well dissolved into hydrophobic solvents such as isooctane. To

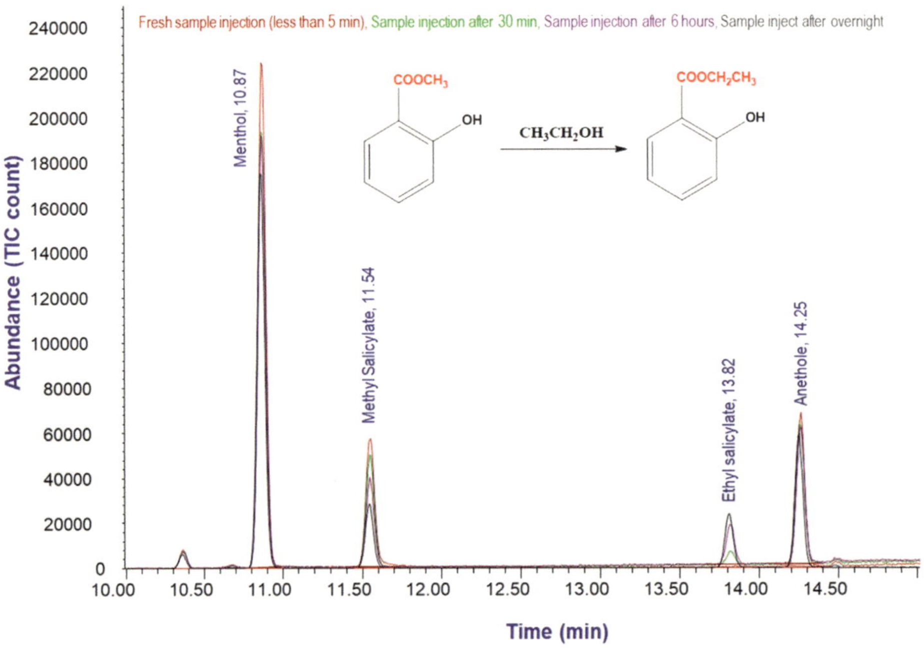

Figure 1.
The GCMS chromatogram of wintergreen flavor extracted by using a water-ethanol solvent system. The peaks of menthol and anethole at the retention times around 10.87 and 14.25 min displayed similar ion count intensity. The methyl salicylate peak intensity at 11.54 min was continuously reduced from brown, green and purple to black after sample extraction, and ethyl salicylate peak intensity at 13.82 min was continuously increased from brown, green, and purple to black. The structure change is demonstrated inside figure.

enhance the extraction efficacy, dispersive liquid-liquid microextraction methods have been reviewed [15]. The emulsion formed between the organic solvent and water can increase the contact surface and enhance the extraction efficiency. Many reports [16–20] have found that the addition of a water-soluble inorganic salt can aid in sufficient dispersion of extraction solvent at microliter level into aqueous phase. Specifically, Na_2SO_4 could enhance the formation of the emulsion between aqueous sample solution and organic solvent [17]. Many studies discovered that a vortex-assisted mixing method helped disperse emulsion and improved extraction efficiency [21–23]. Therefore, an isooctane-water system is commonly used to extract flavor from the toothpaste. Water with Na_2SO_4 was used to disperse and suspend toothpaste matrix into loss slurry at first, and then, isooctane was applied to extract the flavor ingredients. After vortex-assisted mixing, centrifuging at 5000 rpm for 10 min was performed to collect the top isooctane phase for GCMS analysis of flavor ingredients. Most flavor ingredients can be successfully extracted out in this manner, and this method was successful in quantifying menthol and methyl salicylate in the corresponding flavor oils and related products with good recovery and reproducibility (data are not shown here). The advantage of this extraction system is there is no water in the injection solvent, which should improve column life based on the traditional GC column stability consideration. Most modern GC columns except wax and free fatty acid phase (FFAP) columns have significantly improved tolerance to water except when strong acid or base is present in injection liquid. But the presence of water in the injected solvents still exhibited more column bleeding at elevated temperatures. One big disadvantage of this method is that two liquid phases were

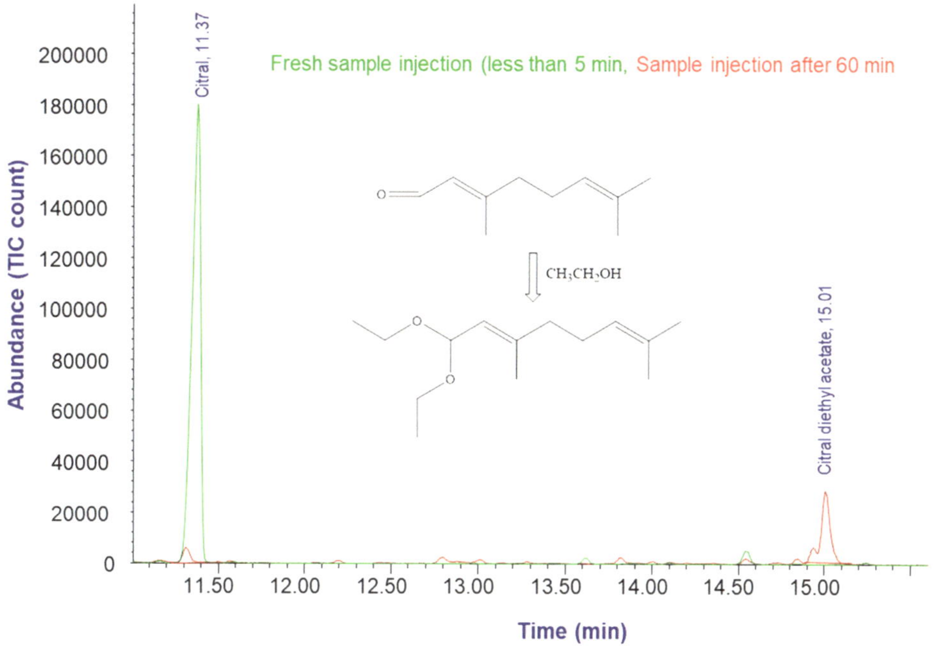

Figure 2.
The GCMS chromatogram of lemongrass flavor extracted by using water-ethanol solvent system. The citral peak intensity at 11.37 min was reduced from green to red after extraction. A new peak at 15.01 min was observed and identified as citral diethyl acetal 60 min after extraction. Both citral and citral diethyl acetal are shown in figure.

involved in the sampling procedures, which could be against the sample homogenization rule in analytical chemistry. The situation is even worse when a full integrated flavor profile is required to evaluate the flavor quality. The flavor ingredients could be unevenly partitioned between those two phases. If only the organic phase is used for injection, this uneven distribution in the two separated liquid phases could twist and change the final analytical results. This phenomenon could be very significant for cleaning products such as toothpastes. Common toothpastes contain about 30% of either glycerol or sorbitol as humectant. Those polyalcohols will mainly stay with water, and they can dramatically enhance the flavor solubility in the water phase and significantly change the flavor ingredient distribution between water and isooctane phases [24]. Polyalcohol impact on flavor distribution during the two-phase fractionation may not be obvious for nonpolar flavor ingredients because they can be well dissolved in hydrophobic isooctane solvent. That is why the quantitation of menthol and methyl salicylate can exhibit a good accuracy and precision result after flavor ingredients were formulated with toothpaste matrix. However, if a full flavor ingredient profile needs to be analyzed to evaluate the flavor quality, this extract system could bring some errors for the final results. Here is an example. When the current procedure was compared with the method described in the Section 3.3 below, the results of toothpaste flavor are shown in **Figure 3**. The two peaks at the retention times of 3.43 and 5.59 min are missing when the current two-phase, water-isooctane, extraction method was applied. Those two peaks were identified as acetic and propanoic acids by using the NIST library and chemical standard. They are the flavors corresponding to sour and cheese characters and commonly used in toothpaste [25]. Those two flavor ingredients are very critical to flavor quality evaluation. Due to their polarity and high

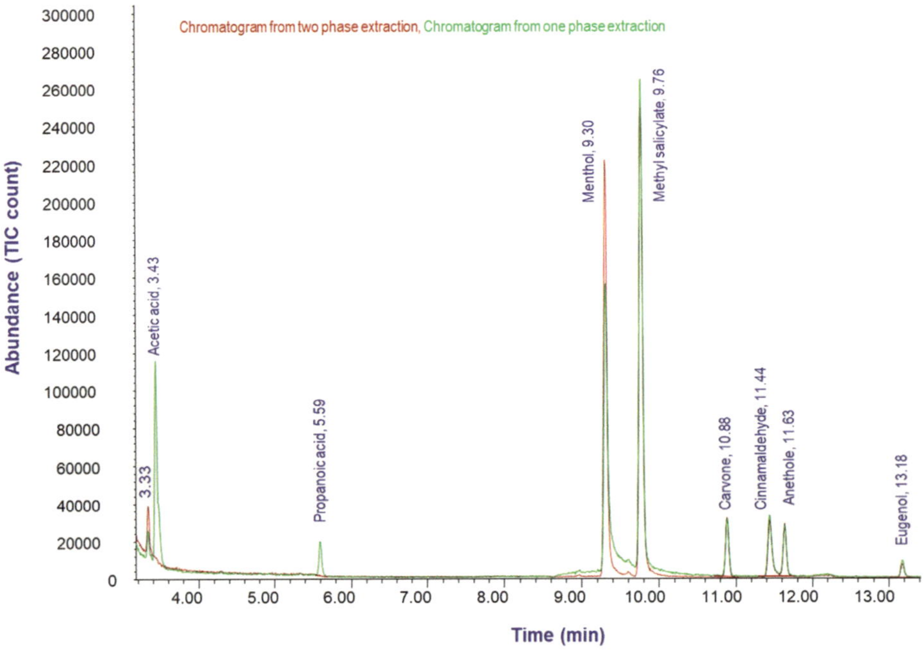

Figure 3.
The GCMS chromatographic flavor profiles from a commercial toothpaste product. The compounds were identified by mass spectra, NIST library, and reference standards. The green chromatogram is extracted by using one phase of water-n-propanol-isooctane (1:8.5:15) solvent system, and the red one is extracted by using two phases of water-isooctane solvent system with fractionation separation.

content of humectants such as glycerol or sorbitol in the water phase, they will stay in the water phase and cannot be injected into the GCMS instrument.

3.3 Sampling procedure with water-n-propanol-isooctane (1:8.5:15) solvent system

To avoid the disadvantage of water-isooctane two-phase application, a new extraction system composed of water-n-propanol-isooctane (1:8.5:15) was explored for flavor ingredient extraction from toothpaste matrix. This single-phase system mimics toothpaste polarity and allows flavor ingredients to be equitably transferred from high viscous paste to light clear extraction solvents. A minimum amount of water was applied to disperse and suspend the toothpaste matrix, and n-propanol was used not only for further dispersion of toothpaste matrix but also to enhance the water solubility in isooctane to avoid phase separation and constitute a one-phase extraction solvent system. Isooctane is a major extraction solvent to get flavor ingredients from the toothpaste matrix. This system was well aligned with the homogenization rule during sampling procedures, which enabled high confidence in detecting all the flavors on the GCMS instrument analysis. By using this ratio of three solvent compositions to mix different commercial toothpaste products, no phase separation was observed and sampling homogenization was achieved during vortex-assisted mixing. This system can dissolve and extract not only hydrophobic flavor ingredients but also hydrophilic flavor ingredients including acetic and propanoic acids, which is demonstrated in **Figure 3** and described in Section 3.2. For a different consumer cleaning product, the ratio of three solvents would need to be slightly adjusted to make sure no phase separation and dominant isooctane is present.

4. Conclusion

This study explored three liquid extraction systems for sampling flavor ingredients from toothpaste products. They are representative liquid sampling procedures to most consumer cleaning products. The first extraction system is the simplest and most eco-friendly[13]. Due to high water activity and nucleophilic attacking capability from the hydroxyl group in ethanol, nucleophilic substitution reaction of ethoxyl from methoxyl group in methyl salicylate compounds and nucleophilic addition reaction from citral to citral diethyl acetal were observed after extraction procedures. In the second system, water-isooctane solvent mixture with two-phase fraction provided a cleaner extraction solution for the GCMS instrumental analysis, especially for hydrophobic flavor ingredient quantitation. However, the ingredient distribution in two phases could be against the homogenization rule to present the holistic sample profiles. Specifically, the high contents of polyalchols such as glycerol and sorbitol present in toothpaste products could prevent polar flavor ingredients such as acetic and propanoic acids from isooctane phase. Therefore, this solvent extraction system is not ideal for flavor profile analysis, and the results could be misleading for flavor quality evaluation. The third liquid extraction system with small amounts of water, suitable amount of n-propanol and large amounts of isooctane provided a one-phase extraction media after mixing with the toothpaste products, which can better satisfy the sampling homogenization rule in the analytical chemistry. This system is not only suitable for flavor ingredient quantification but also can be used for toothpaste product flavor profiling analysis.

Conflict of interest

The authors declare no conflict of interest statement.

Author details

Zhigang Hao*, Vivian Liu, Jake Salerno, Yu Wang, Mania Bankova and Long Pan
Global Technology Center, Colgate-Palmolive Company, Piscataway, NJ, USA

*Address all correspondence to: zhigang.hao@gmail.com

References

[1] Wilkes JG, Conte ED, Kim Y, Holcomb M, Sutherland JB, Miller DW. Sample preparation for the analysis of flavors and off-flavors in foods. Journal of Chromatography. A. 2000;**880**(1-2):3-33. DOI: 10.1016/S0021-9673(00)00318-6

[2] Zellner BA, Dugo P, Dugo G, Mondello L. Gas chromatography–olfactometry in food flavour analysis. Journal of Chromatography. A. 2008;**1186**(1-2):123-143. DOI: 10.1016/j.chroma.2007.09.006

[3] Almeida SAAG, Meneses AC, Araújo PHH, Oliveira D. A review on enzymatic synthesis of aromatic esters used as flavor ingredients for food, cosmetics and pharmaceuticals industries. Trends in Food Science & Technology. 2017;**69**:95-105. DOI: 10.1016/j.tifs.2017.09.004

[4] Smith RM. Before the injection-modern methods of sample preparation for separation techniques. Journal of Chromatography. A. 2003;**1000**(1-2):3-27. DOI: 10.1016/S0021-9673(03)00511-9

[5] Sasidharan S, Chen Y, Saravanan D, Sundram KM, Latha YL. Extraction, isolation and characterization of bioactive compounds from plants' extracts. African Journal of Traditional, Complementary, and Alternative Medicines. 2011;**8**(1):1-10. DOI: PMCID: PMC3218439

[6] Azmir J, Zaidul ISM, Rahman MM, Sharif KM, Mohamed A, Sahena F, et al. Techniques for extraction of bioactive compounds from plant materials: A review. Journal of Food Engineering. 2013;**117**:426-436. DOI: DOI; 10.1016/j.jfoodeng.2013.01.014

[7] Hennion MC, Cau-Dit-Coumes C, Pichon V. Trace analysis of polar organic pollutants in aqueous samples: Tools for the rapid prediction and optimization of the solid-phase extraction parameters. Journal of Chromatography. A. 1998;**823**(1-2):147-161. DOI: 10.1016/S0021-9673(98)00479-8

[8] Majors RE. An overview of sample preparation methods for solids. LC-GC European. 1999;**17**(6):8-13

[9] Berendsen BJA, Wegh RS, Meijer T, Nielen MWF. The assessment of selectivity in different quadrupole-orbitrap mass spectrometry acquisition modes. Journal of the American Society for Mass Spectrometry. 2015;**26**(2):337-346. DOI: 10.1007/s13361-014-1021-x

[10] Poole SK, Dean TA, Oudsema JW, Poole CF. Sample preparation for chromatographic separations: An overview. Analytica Chimica Acta. 1990;**236**(1):3-42. DOI: 10.1016/S0003-2670(00)83297-3

[11] Hernández Y, Lobo MG, González M. Factors affecting sample extraction in the liquid chromatographic determination of organic acids in papaya and pineapple. Food Chemistry. 2009;**114**:734-741. DOI: 10.1016/j.foodchem.2008.10.021

[12] Torssell KG. Natural Product Chemistry: A Mechanistic, Biosynthetic and Ecological Approach. 1st ed. Stockholm, Sweden: Swedish Pharmaceutical Press; 1997. pp. 12-40

[13] Fu Y, Hao Z, Parker B, Knapp M. A greener way to screen toothpaste for diethylene glycol. American Journal of Analytical Chemistry. 2011;**2**:938-943. DOI: 10.4236/ajac.2011.28109

[14] FDA. 2007. https://www.fda.gov/Food/FoodScienceResearch/

LaboratoryMethods/ucm113209.htm. [Accessed: May 31, 2022]

[15] Li MJ, Zhang HY, Liu XZ, Cui CY, Shi ZH. Progress of extraction solvent dispersion strategies for dispersive liquid-liquid microextraction. Chinese Journal of Analytical Chemistry. 2015;**43**(8):1231-1240. DOI: 10.1016/S1872-2040(15)60851-9

[16] Hosseini MH, Rezaee M, Mashayekhi HA, Akbarian S, Mizani F, Pourjavid MR. Determination of polycyclic aromatic hydrocarbons in soil samples using flotation-assisted homogeneous liquid-liquid microextraction. Journal of Chromatography. A. 2012;**1265**:52-56. DOI: 10.1016/j.chroma.2012.09.099

[17] Ma H, Li Y, Zhang H, Shah SM, Chen J. Salt-assisted dispersive liquid–liquid microextraction coupled with programmed temperature vaporization gas chromatography–mass spectrometry for the determination of haloacetonitriles in drinking water. Journal of Chromatography. A. 2014;**1358**:14-19. DOI: 10.1016/j.chroma.2014.06.021

[18] Hosseimi MH, Asaadi P, Rezaee M, Reza Rezaei M, Reza Pourjavid M, Arabieh M, et al. Homogeneous liquid-liquid microextraction via flotation assistance (HLLME-FA) method for the pretreatment of organochlorine pesticides in aqueous samples and determination by GC–MS. Chromatographia. 2013;**76**:1779-1784. DOI: 10.1007/s10337-013-2570-z

[19] Moniruzzaman M, Rodriguez I, Rodriguez-Cabo T, Cela R, Sulaiman SA, Gan SH. Assessment of dispersive liquid-liquid microextraction conditions for gas chromatography time-of-flight mass spectrometry identification of organic compounds in honey. Journal of Chromatography. A. 2014;**1368**:26-36. DOI: 10.1016/j.chroma.2014.09.057

[20] Haddadi H, Rezaee M, Semnani A, Mashayekhi HA. Application of homogeneous liquid–liquid microextraction via flotation assistance followed by gas chromatography for determination of butachlor in water samples. Bulletin of the Chemical Society of Japan. 2013;**86**:1282-1286. DOI: 10.1246/bcsj.20130175

[21] Gure A, Lara FJ, Garcia-Campana AM, Megersa N, Olmo-Iruela MD. Vortex-assisted ionic liquid dispersive liquid-liquid microextraction for the determination of sulfonylurea herbicides in wine samples by capillary high-performance liquid chromatography. Food Chemistry. 2015;**170**:348-353. DOI: 10.1016/j.foodchem.2014.08.065

[22] Seebunrueng K, Santaladchaiyakit Y, Srijaranai S. Vortex-assisted low density solvent based demulsified dispersive liquid–liquid microextraction and high-performance liquid chromatography for the determination of organophosphorus pesticides in water samples. Chem. 2014;**103**:51-58. DOI: 10.1016/j.chemosphere.2013.11.024

[23] Seebunrueng K, Santaladchaiyakit Y, Srijaranai S. Vortex-assisted low density solvent liquid–liquid microextraction and salt-induced demulsification coupled to high performance liquid chromatography for the determination of five organophosphorus pesticide residues in fruits. Talanta. 2015;**132**:769-774. DOI: 10.1016/j.talanta.2014.10.020

[24] Scheepers JJ, Muzenda E. Glycerol—A viable solvent for absorption of highly polar solutes I: Behaviour of molecular interactions. Journal of Clean Energy Technology. 2015;**3**(4):282-286. DOI: 10.7763/JOCET.2015.V3.209

[25] Liu C, Yang Q, Linforth R, Fisk ID, Yang N. Modifying Robusta coffee aroma by green bean chemical pre-treatment. Food Chemistry. 2019;**272**:251-257. DOI: 10.1016/j.foodchem.2018.07.226

Chapter 5

Recent Applications of Gas Chromatography in Bioanalysis

Victor David and Serban C. Moldoveanu

Abstract

Bioanalysis involves a broad range of chemical analyzes. These analyzes include that of biotics, such as natural components of living organisms, as well as xenobiotics, such as drugs and their metabolites in biological systems. Because many biotics and xenobiotics are not volatile molecules, the main technique for bioanalysis is high-performance liquid chromatography (HPLC) and the limitation of GC utilization is caused by the fact that GC is applicable only to volatile samples. However, gas chromatography (GC) in particular coupled with mass spectrometry (MS) as detection is also a very useful technique in bioanalysis. A considerable number of analytes in bioanalysis are volatile or can be made volatile following, for example, derivatization. As a result, GC (and GC/MS) are commonly utilized for the analysis of biotics, such as amino acids, fatty acids, various metabolites in biological fluids, and in particular of a large number of xenobiotics, such as drugs, drug metabolites, toxicants, and certain metabolic compounds caused by toxicants. The chapter will present progress in the GC methodology for extending its applicability to bioanalysis and will provide a review of more recent applications.

Keywords: bioanalysis, gas chromatography, mass spectrometry, volatile compounds, toxicants, biotics, biomatrix

1. Introduction

Biological samples include biotics, which are natural components of living organisms, as well as xenobiotics, such as drugs and their metabolites, toxicants, certain metabolic products caused by toxicants, and other components of biological systems. Biometrics may consist of various body tissues, blood, plasma, serum, hair, milk, saliva, sweat and skin surface lipids, urine, fecal materials, and breath. Biological samples typically have a very complex composition and provenience. Bioanalysis involves a broad range of chemical analyzes and chromatographic techniques are the ones most frequently utilized for this purpose. The most common chromatographic technique used in bioanalysis is high-performance liquid chromatography (HPLC), but gas chromatography (GC) is also frequently utilized. The present chapter describes some of the more recent applications of GC in bioanalysis.

2. Short overview of gas chromatography

Gas chromatography (GC) is one of the main types of chromatographic techniques that has gas as the mobile phase, usually helium or hydrogen. As an analytical technique, the separation by GC is always coupled with a detection technique. Gas chromatography requires the sample to be in gas form during the separation, and this causes its limitation to the analysis of only volatile compounds. However, many nonvolatile compounds can be transformed into volatile compounds by chemical modifications (derivatization).

The separation in GC takes place in a chromatographic column, the modern instruments use open-tubular columns (capillary columns), while packed columns that were used in the past are now much less utilized. The capillary column, commonly made from silica, has an inner coating with a film that acts as a stationary phase that can have different chemical structures. The stationary phase film is selected based on the intended separation, and a variety of such phases are commercially available. The chromatographic column is placed in an oven with a controlled temperature kept constant during the separation process (isocratic conditions) or modified following a specific program (gradient conditions). The sample to be analyzed is introduced at the head of the chromatographic column using an injector. For liquid samples, the injector typically uses a syringe placing a precise volume, such as 0.5, 1.0, or 2.0 μL, into the injection port of the GC that is heated at a specific temperature, volatizes the liquid, and places the sample in the gas flow of the chromatographic column. The gas samples are usually placed in the gas flow of the GC using a loop of specific volume that is connected to the gas flow of the instrument. Various solvent-less type injections are also possible [1]. The separated components of the sample are carried into a detector that generates an electric signal. The signal is proportional to the instantaneous concentration (or amount) of the analyte passing through the detector, allowing the use of the signal for quantitative measurements. Several types of detectors are used in GC. The most common detectors in GC are probably the flame ionization detector (FID) and the mass spectrometer (MS). Other detectors, such as thermal conductivity (TCD), nitrogen phosphorus (NPD), electron capture (ECD), photoionization (PID), are also used in GC. Some of these detectors have a universal response to the analytes and provide only quantitative information (e.g., FID), other detectors are element specific (e.g., NPD), and others offer both the capability of quantitative and qualitative identification (e.g., MS). The MS detector generates a total signal that can be used for quantitation, but also mass spectra for the compounds passing the detector. Large libraries of mass spectra (some libraries containing spectra for over 800,000 compounds) allow qualitative identification of unknown components in the sample, or confirmation of the nature of the evaluated analyte. The detectors have different sensitivities and linear ranges, this depending on the type of detector but also on the detector manufacturer.

3. Gas versus liquid chromatography in bioanalysis

Because of the common complexity of biological samples, bioanalyses usually include separations that are carried out with chromatographic techniques [2–4]. These analyses produce quantitative and/or qualitative information about biotics, as well as xenobiotics. A biomatrix typically consists of three main components: large molecules, such as proteins, small (non-polymeric) organic molecules that are typically the target for analysis in GC, and inorganic content. Bioanalytical investigations are commonly based on a protocol that should be focused on two aspects, namely, the sample preparation and chromatographic methodology.

Although high-performance liquid chromatography (HPLC) can be considered as the technique most commonly utilized in bioanalysis due to the more options to investigate the complex biological matrix and the more large area of compound amenable to the analysis by this technique, GC in spite of its smaller covered domain of applications has several advantages, such as higher separation capacity, excellent sensitivity, and much better capability of unknown compound identification (when coupled with MS). Besides sample enrichment and interferences removal from the biological matrix, which is common to both GC and HPLC [5, 6], sample preparation for GC can include the role of transforming the nonvolatile analytes in volatile compounds. This can be performed by various techniques and can use a range of derivatization reagents [7–10]. Generally, the sample preparation is carried out by extraction techniques, including liquid-liquid, supercritical fluid, solid-phase, or microextractions, which may include or not a derivatization step for improving volatility, detectability, or improvement of separation capability [8]. In conclusion, sample preparation for GC analysis has the main role of sample simplification, analyte concentration, or analyte structure modification (derivatization). Derivatization in GC can be used to render the analytes more volatile and thermally stable, to improve separation, detectability, and accuracy of analysis.

4. GC determination of volatile organic compounds of biological origin

Usually, a volatile organic compound (VOC) is characterized by a minimum 0.13 kPa vapor pressure at 20°C [11] and can belong to various classes, such as aliphatic and aromatic hydrocarbons, alcohols, amines, ketones, aldehydes, acids and their derivatives, sulfur compounds, and many compounds with multiple functionalities. Volatile organic compounds (VOC) of biological origin (e.g., human) are chemical components from breath [12], as well as volatiles emitted from the skin, or bodily fluids (urine and feces), many of them being odorous and different from the metabolites produced in the axillae (underarms) [13]. Identification of these compounds and their concentrations could be useful for assessing various diseases that include pulmonary diseases, liver dysfunctions, kidney diseases [14], gastrointestinal problems, diabetes, and others. Therefore, the investigation of the content of such samples by various analytical techniques, including GC, can be considered a potentially noninvasive means of diagnosis, monitoring of pathological processes, and assessment of pharmacological response, being fast, simple, and acceptable to patients [15]. However, unlike solid or liquid biomatrices, gaseous samples are more difficult to be sampled and sampling is critical in the analytical process.

Gaseous samples of biological provenience can be collected by different techniques, including solid-phase microextraction (SPME) [16], adsorption on graphitized carbon, molecular sieve, various resins (e.g., XAD resins), or by condensation in cold traps (cryfocusing), when the water should be removed selectively before analysis [17]. Also, a sampler can be a cylinder that can be placed on the skin surface in order to create a headspace [18]. In the case of using SPME for breath sampling, this approach should avoid the collection of droplet phases of exhaled breath by using a filter-incorporated needle-trap device [19]. Needle-trap devices are recommended for the extraction of VOCs from biological solid and liquid samples. They are capable of satisfying many actual demands, as green analytical methods, eliminating solvent consumption, and performing an on-site sampling. After extraction of VOCs, they are thermally desorbed and automatically injected into the GC system for separation and quantitation [20]. Specifically for breath analysis, sampling and storage can also be chosen from several modalities, such as polymeric sampling bags (Tedlar®, Nalophan, Cali-5-Bond), syringes, gas evacuated steel, or glass containers [21]. Sample collection can be performed offline, preserving stability in a time of the samples, or online, which is less used in practice [22].

Analytical methodologies based on GC with mass spectrometry (MS) have been developed for the determination of exhaled VOCs pattern and have become a potential method for early diagnosis of lung cancer. Thus, GC-MS analysis for many patients revealed that 1-butanol and 3-hydroxy-2-butanone are biomarkers that at significantly higher concentrations in breath can diagnose lung cancer [23]. There are other studies that reported the analysis of exhaled volatile carbonyl compounds for the identification of specific carbonyl cancer markers to differentiate benign pulmonary disease from early-stage lung cancer and to compare its diagnostic accuracy with positron emission tomography scans [24]. Asthma can be diagnosed by measuring the concentration levels of ethane, pentane, 8-isoprostane, and NO [25].

Other examples of VOCs as biomarkers determined by GC technique and their concentrations used for illness diagnostics include, for example, (i) alkanes, monomethylated alkanes for breast cancer; (ii) S-containing compounds, such as methyl mercaptan, dimethylsulfide for hepatic coma and cholera [15, 26]; (iii) carbon disulfide, pentane, ethane for schizophrenia; (iv) hexanal, 1-octen-3-ol and octane for liver cancer [27]; and (v) nonanal, 2-ethylhexan-1-ol, 5-ethyl-3-methyloxolan-2-one, heptan-2-one, 1,1,4a-trimethyl-4,5,6,7-tetrahydro-3H-naphthalen-2-one and propan-2-one from urine as surveillance biomarker of bladder cancer [28].

One of the common possibilities to diagnose gastrointestinal disease relies on the analysis of VOCs from fecal materials. For example, the analysis of several VOCs from chicken feces with and without *Campylobacter jejuni* revealed the abundance of six VOCs, considered fecal biomarker for this bacteria in chicken feces, namely, hexanal, 2-octenal, pyrrole, ethyl acetate, methanol, and 2-heptanone [29].

5. GC analysis of biotics in biological fluids

There is a large variety of chemical species as part of the biomatrices. Among them, amine-type compounds, including polyamines, amino acids, catecholamines, fatty acids, and carbohydrates (monosaccharides, disaccharides, oligosaccharides, and polysaccharides), are essential for cell metabolism. These compounds are very polar and many of them are lacking a chromophore moiety in order to be detected by UV–Vis spectrometry [30].

The determination of amino acids in biological fluids can serve as cancer biomarkers [31, 32]. The first choice in selecting chromatographic techniques is liquid chromatography, which offers various retention mechanisms for amino acids [33], but gas chromatography coupled with mass spectrometry is still being used for the analysis of these compounds from biomatrices due to its advantage in compound identification. The difficulty in GC determination of these very polar compounds is their lack of volatility, and the only possibility to use this chromatographic technique is to apply a derivatization procedure in order to decrease their polar character and transform amino acids and more volatile species. The derivatization procedure can take place simultaneously with an extraction operation, or can be performed separately, before extraction [34]. Besides isolating the derivatives from an aqueous medium, the extraction has the role of eliminating part of the sample matrix, and in some cases the enrichment of the target analytes [8].

Derivatization of amino acids with alkyl chloroformates, such as propyl chloroformate [35], is commonly used and the derivatization can be carried out directly in the biological samples that do not need prior protein precipitation or solid phase extraction of the amino acids. The reaction takes place rapidly (less than 1 min) in a water/propanol/pyridine medium. The analytes are further extracted in chloroform and analyzed by GC or GC-MS. Other alkyl chloroformates can be used for the same purpose [36]. The derivatization at both functional groups, amino, and carboxyl, of amino acids, will lead to stable volatile derivatives that are further extracted in an organic solvent (e.g., chloroform or isooctane) and injected into the GC-MS system. By these GC-MS methods, a large number of amino acids and dipeptides were determined allowing limits of detection (LOD) situated in the range of 0.03–12 μmoles/L [37].

An older study compared the main derivatization reagents used for GC analysis of amino acids in complex samples (lyophilized *E. coli* microbial culture) [38]. These reagents were N,O-bis(trimethylsilyl)trifluoroacetamide (BSTFA), N-methyl-N-(*tert*-butyldimethylsilyl) trifluoroacetamide (MTBSTFA), and isobutyl chloroformate (iBuCF). The study showed that the performances in the case of silylation with the aid of MTBSTFA are comparable to those obtained for acylation with iBuCF, but require a more laborious extraction procedure to isolate the amino acids prior to derivatization, and determination of amino acids as N(O,S)-alkoxycarbonyl alkyl esters is more advantageous for this type of application.

Derivatization with BSTFA of amino acids, attaching trimethylsilyl at both functional groups, was applied for their analysis in urine and cerebrospinal fluid collected from rats [39]. The derivatives of amino acids were isolated with the aid of hollow fiber solid phase microextraction (HF-SPME) and were analyzed by GC-MS. This method generated limits of detection (LOD) situated in the range of 0.3–17 ng/mL [40]. A version of this derivatization reaction, assisted by microwave, was used for the determination of the concentrations of L-amino acids in cerebrospinal fluid in order to have a biochemical insight into central nervous system disorders [41]. A review on this topic has been very recently published and offers information on new potential biomarkers of central nervous system diseases investigated by GC-MS [42].

A variety of fatty acids play critical roles in biological systems. They exist in the diet of humans, in the bloodstream, cells and tissues of humans, both being an energy sources and membrane constituents. GC with different detections is the most widely used analytical technique for the separation and quantitation of fatty acids as methyl ester derivatives. Complete separation of common fatty acids is currently carried out by using capillary columns with polar stationary phases, and the use of a flame ionization detector (FID) offers sufficient sensitivity to measure them from complex samples. In specific applications, depending on the class of lipids to be separated, further separation and fractionation should be necessary or different columns required for GC separations [43].

Fatty acids can influence cell and tissue metabolism, function, and responsiveness to hormonal and other signals, and imbalances in fatty acids are related to a variety of diseases, which makes the measurement of fatty acids in biological samples very important [44]. For example, fatty acids (octadecanoic acid, heptadecanoic acid, tetradecanoic acid, eicosanoic acid, and cis-vaccenic acid) and their esters showed altered levels in breast cancer patients in several studies, and the oxidation of these acids is important in the development of tumor cells [45]. The use of GC-MS-based investigations allowed the identification of several metabolites resulting from metabolic processes [46, 47]. Some of the main strategies include extraction methods (e.g., liquid-liquid extraction and solid-phase microextraction), derivatization methods, column selections, and internal standard selections, in order to identify and measure the concentration of various fatty acids in biomatrices. A recent GC-MS analytical procedure for the rapid and selective derivatization of free fatty acids into methyl esters directly in plasma without transmethylation of lipid-bound fatty acids was developed based on the reaction with CH_3I in dimethyl sulfoxide and in the presence of solid bases (sodium carbonate) [48]. The method requires a very small volume of plasma (50 μL) and has a detection limit of 0.1 ng/mL. The relationships between fatty acid imbalances and the investigated diseases [44], or the influence of the use of drugs [49] have been recently reviewed.

Lipids are also frequently analyzed using GC. A variety of lipids are present in living organisms, such as glycerolipids, sterols, stanols, prenols, and phosphoglycerides. Although the direct analysis of lipids is difficult using GC, the technique is commonly used for the analysis of fatty acids present in lipids and of other lipid components [50]. Some lipids can be directly analyzed using GC-MS [51], but hydrolysis of lipids and derivatization of the fatty acids is a common procedure for analysis using GC separation.

Determinations of carbohydrates by GC are limited to mono-, di- and trisaccharides, and these can be performed after derivatization to enhance their volatility and thermal stability [46]. Upper saccharides are less stable at elevated temperatures used in GC, and only a few applications are known for these compounds. Generally, the lower saccharides are transformed into methyl ethers, acetate, trifluoroacetate, and trimethylsilyl derivatives that are separated and detected by GC.

Silylation reagents used for carbohydrate analysis include hexamethyldisilazane (HMDS), trimethylchlorosilane (TMCS), N-trimethylsilylimidazole (TMSI), N-methyl-N-trimethylsilylacetamide (MSA), N-trimethylsilyldiethylamine (TMSDEA), N-trimethylsilyldimethylamine (TMSDMA), N-methyl-N-trimethylsilyltrifluoroacetamide (MSTFA), N,O-bis(trimethylsilyl)acetamide (BSA) and N,O-bis(trimethylsilyl)trifluoroacetamide (BSTFA) [52]. The derivation conditions depend on the type of reagent and samples to be analyzed. Comparison of the analytical performances of GC-MS based on silylation derivatization with other derivatization procedures based on other chromatographic techniques, mainly with hydrophilic interaction liquid chromatography (HILIC), reversed-phase liquid chromatography (RP-LC) applied to biological samples has been recently reported [53].

From the class of derivatization of carbohydrates by alkylation reactions, the most convenient is permethylation. In practice, this reaction can be achieved using CH_3I, in the presence of dimethyl sulfoxide and a solid base (NaOH, KOH, and potassium tert-butoxide) [54, 55].

The alkylation of sugars can be performed in only one step by adding dimethyl sulfoxide, powdered sodium hydroxide, and methyl iodide directly to an aqueous solution of the sample. This procedure can be applied also to aqueous samples by an additional excess of sodium hydroxide [56]. The procedure has been applied to many biological systems, for example, ref. [57–59].

A few analytical studies have been dedicated to the determination of fatty alcohols, attempting to elucidate the role of fatty alcohols in biological systems which is still uncertain. So far, it is known that an inherited disorder of fatty alcohols metabolism, known as Sjögren-Larsson syndrome, is a consequence of a deficiency in fatty alcohols oxidation. A report describing the determination of these compounds in rat plasma samples is known, based on derivatization with pentafluorobenzoyl chloride and gas chromatography/electron capture negative ion chemical ionization-MS [60].

6. GC analysis of drugs and metabolites in main biological matrices

Drugs and, in many cases, their metabolites are of interest of being analyzed in biomatrices for clinical and pharmacological purposes. Biological samples are frequently processed in the view of GC analysis. The processing can be achieved by several liquid-liquid or solid-phase extraction techniques. This part of sample preparation has the role of simplification of sample composition and in several circumstances to enrich the sample. The main purpose of sample preparation is however to enhance the volatility of target compounds, drugs or their metabolites, which are nonvolatile in many cases. Therefore, this task is crucial for the performance of GC analysis, the method is usually applied to a large number of samples [5]. Derivatization can take place before, simultaneously with or after the extraction. This process is dependent on the detection used in GC analysis. Validation is another important part of the GC protocol and should take into consideration all aspects of the analytical process based on GC separation and detection [61]. The criteria for GC methods have not changed too much in time and they include the proof for stability, selectivity, accuracy, precision (intra- and inter-day), recovery, response linearity, computation of detection limit (LOD) and quantitation limit (LOQ), ruggedness, and other issues important to be applied to a large number of biosamples. The protocol for validation may also require re-validation, cross-validation, endogenous drug evaluation, and evaluation of matrix effects [62].

The derivatization method is usually chosen from this list of reactions: silylation, alkylation, acylation, and the formation of cyclic derivatives. A long list of derivatization reagents is available for these applications, but in practice, methodologies based on the several formations of trimethylsilyl, perfluoroacyl, or methylated-derivatives have proved to be the most versatile and extensively used [63]. The use of GC-MS systems for bioanalysis is almost always recommended for structural confirmation, and this is facilitated by the existence of comprehensive libraries containing reference MS spectra for different derivatives of many drugs and their metabolites [63]. Although much less utilized, GC coupled to Fourier transform infrared spectroscopy (FTIR) is also an alternative technique to GC-MS, providing structural information that allows, for example, the discrimination between isobars and isomers [64].

The literature reports several reviews focused on GC applications for bioanalysis for specific classes of drugs and their metabolites [65–67]. Some examples of analytical methods based on GC techniques are listed in **Table 1**. The majority of these techniques are applied to detect drugs of abuse, such as opiates, cocaine, cannabis, amphetamines, or benzodiazepines.

Various technical aspects of using GC/MS analysis in bioanalysis are related to the complexity of the matrix of the samples of biological origin. For example, in using GC/MS on biological samples one problem is the possibility that the matrix is influencing the intensity of MS signals by the co-eluting species with the target analytes. Interferences effects are characterized by signal enhancement or suppression [81]. This phenomenon is known as the matrix effect and can be observed also in liquid chromatography coupled to electrospray ionization MS. In GC-MS the matrix effect is usually less important compared to LC-MS or LC/LC-MS. For this reason, in GC-MS

Drug name	Biological matrix	Sample preparation	GC details	Analytical performances	Ref.
Opioids: codeine, morphine, heroin, 6-acetylmorphine, desomorphine, ethylmorphine, methadone, 2-ethylidene-1,5-dimethyl-3,3-diphenylpyrrolidine, 2-ethyl-5-methyl-3,3- diphenyl-1-pyrroline, papaverine, tramadol, O-desmetyltramadol, tapentadol	Whole blood	QuEChERS method consisted of the pretreatment of the whole blood samples using ultrasonication, the use of ethyl acetate as extraction solvent, and a previous step of sample alkalinization; phenacetin as internal standard (IS).	Capillary column (30 m; 0.25 mm; 0.25 μm, cross-linked 5% diphenyl and 95% dimethyl polysiloxane); splitless injection; MS detection in ion monitoring mode (SIM).	Linearity between 31 and 2000 ng/mL;	[68]
Glimepiride	Urine, tissues sample (a portion of the kidney, liver, spleen, and intestine)	Liquid-liquid extraction method was employed by using 1-butanol: hexane (50:50, v/v); derivatization employing N-methyl-N-(trimethylsilyl) trifluoroacetamide	30m × 0.32 mm, 0.25 μm, and 1,4-bis (dimethylsiloxy)phenylene dimethylpolysiloxane column; MS detection in TIC mode	Linearity between 500 to 2500 ng/ml	[69]
Cocaine (metabolites), methadone, and morphine	Postmortem adipose tissue	Aqueous acid extraction of analytes, alkalinization of the extract, solid-phase extraction + elution with $CHCl_3$, and derivatization with BSTFA.	Capillary column EVDX-5MS 5% PH ME Siloxane (12.5 m × 0.20-mm i.d., 0.33-μm film thickness); Deuterated compounds are used as internal standards.	Linearity between 0.1 to 1.000 μg/g; Limits of detection: – 0.005 μg/g for cocaine cocaethylene and methadone, – 0.02 μg/g for benzoylecgonine, – 0.01 μg/g for ecgonine methyl ester and morphine	[70]
Clozapine	Human plasma	Dispersive liquid-liquid microextraction with $CHCl_3$, and derivatization with trifluoroacetic anhydride (TFAA).	Capillary column 30m x 250 μm i.d. internally coated with 0.35 μm thick film of 5%-(phenyl) methylpolysiloxane, MS detection in SIM (selected ion monitoring) mode; Clozapine-d4 as IS.	Linearity domain over 50–800 ng/mL; Limit of quantification was set at 50 ng/mL	[71]

Drug name	Biological matrix	Sample preparation	GC details	Analytical performances	Ref.
Cannabinoids: cannabidiol (CBD); tetrahydrocannabinol (THC) cannabinol (CBN)	Hair samples from patients	Extraction from hair in NaOH solution + Solid-phase extraction followed by derivatization with MSTFA	No detail on GC separation; MS detection in multiple reaction-monitoring (MRM) mode; deuterated internal standards: CBD-D3, THC-D3, and CBN-D3	CBD concentrations ranged from 10 to 325 pg/mg of hair	[72]
Prednisolone, prednisone, cortisol, and cortisone	Plasma samples from patients with nephrotic syndrome during oral prednisolone therapy	SPE on Sep-Pak C18 Plus short-body cartridge; derivatization with heptafluoro-*n*-butyric anhydride (HFBA).	Capillary GC with SPB-1 fused silica capillary column (15 m length × 0.25 mm i.d.) with 0.25 μm film thickness of stationary phase; MS detection in SIM mode	Linearity between 10 and 500 ng/mL for prednisolone; 10 and 115 ng/mL for prednisone; 1 and 140 ng/mL for cortisol; 1 and 65 ng/mL for cortisone.	[73]
Alfentanil; fentanyl and sufentanil	Urine and plasma	Hollow fiber liquid-phase microextraction using hexyl acetate as extraction solvent	DB–35MS, 30 m×0.25 mm, capillary column with a 0.15 μm stationary phase thickness; nitrogen phosphorus detection (NPD)	LODs between f 8 and 15 ng/L.	[74]
Amphetamine-type stimulants: amphetamine, methamphetamine, para-methoxyamphetamine, and (±)-3,4-methylenedioxy methamphetamine); synthetic cathinones: mephedrone, buphedrine (buphedrone ephedrine metabolite), 4-methylephedrine (mephedrone metabolite), and pentylone).	Urine samples	SPME with fiber tips C18, C18-SCX (mixed mode), and PDMS-DVB; derivatization with pentafluoropropionic anhydride (PFPA) in ethyl acetate.	DB-5ms (5% phenyl/95% methylpolysiloxane); 30 m x 0.25 mm, 0.25 μm thickness) column; MS detection in selected ion monitoring (SIM) mode; Deuterated internal standards	Limits of detection (LOD) between 5 and 25 ng/mL; Low limits of quantification (LLOQ) between 25 and 100 ng/mL.	[75]
Cathinone-type synthetic drugs: 4-fluoromethcathinone, methcathinone, 4-methylethcathinone, 3,4-dimethylmethcathinone 4-ethylmethcathinone	Human urine	The two-step derivatization: (1) oximation with hydroxylamine hydrochloride; and (2) trimethylsilylation with MSTFA.	Capillary GC column: BPX5, 30 m × 0.25 mm; 0.25 μm thickness; MS detection.	Limit of quantitation (LOQ): 15–24 μg/mL.	[76]

Drug name	Biological matrix	Sample preparation	GC details	Analytical performances	Ref.
Catecholamine metabolite: 3,4-dihydroxyphenylglycol	Human urine patients suffering from chronic inflammatory rheumatic diseases	Extraction in ethyl acetate and derivatization with pentafluorobenzyl bromide.	DB-5ht fused silica column (15 m × 0.25 mm i.d., 0.1μ m film thickness); MS with electron-capture negative-ion chemical ionization; trideutero 3,4-dihydroxyphenylglycol	LOD: 76 amol	[77]
Carvedilol	Human plasma; bioequivalence study.	Extraction with diethyl ether and ethyl acetate and derivatization with MSTFA.	Separation with HP-5 MS column with 0.25 μm film thickness; 30 m length × 0.25 mm i.d.; MS detection	Linearity between 15 and 500 ng/mL; LOD = 5 ng/mL, and LOQ = 15 ng/mL	[78]
Diclofenac	Human plasma; bioequivalence study.	Extraction in hexane; derivatization with pentafluoropropionic anhydride (PFPA).	BP-1 fused silica capillary column (15 m × 250 μm × 0.25 μm); MS in SIM mode; 4-hydroxydiclofenac as internal standard	Linearity between 0.25 and 50 ng/mL; LOD = 0.125 and LOQ = 0.25 ng/mL	[79]
Metformin	Human serum and urine	Derivatization with pentafluoropropionic anhydride in ethyl acetate; extraction with toluene.	Capillary column Optima 17 (15 m x 0.25 mm I.D. 0.25 μm film thickness); MS detection in negative-ion chemical ionization; metformin-d6 as internal standard.	Detection limit of 300 fmol	[80]

Table 1.
GC methods for the determination of drugs in biological matrices.

the possibility of interference is frequently neglected in spite of the fact that it still should be taken into consideration when sample preparation does not remove entirely the sample matrix. The difference between the effect of matrix on the two chromatographic techniques HPLC-MS and GC-MS is caused by the fact that the two methods are based on different ionization mechanisms, and the matrix effects can have different intensities in affecting ionization. In HPLC-MS the ionization mechanism is a soft process in the interface of MS, while in GC-MS this process involves higher energy that overcomes the competition with the other possible co-eluted compounds. One of the solutions to compensate for the interference effects is the use of stable isotopically labeled standards of the target analytes [82, 83], such that both the analytes and the standard are equally affected by interferences.

7. GC determination of toxicants in biological matrices

Forensic toxicology deals with the investigation of all substances of exogenous origin that do not have a normal physiological role in the biochemical processes of the organisms [84, 85]. This includes analytical toxicology, which is focused on methodologies for the identification and quantitation of chemical substances that have adverse effects on living organisms. These analyses in biomatrices are very important, and frequently the interest is directed toward the analytical investigation of toxic species in various consumable products (e.g., food, beverages, nutraceuticals, agricultural products, pharmaceuticals, environment, or tobacco products), and also to the detection of various metabolites of these species in biological fluids and living organisms. The domain of concentrations that need to be determined by GC is generally very broad and depends on the sample provenience and the degree of contamination of investigated samples. Some recent applications of the use of GC-MS in this domain are further presented.

GC-MS techniques can be utilized, for example, for the evaluation of certain metabolic disruptions caused by various toxicants [86]. Thus, the high sensitivity of GC-MS was useful for investigating the effect of different concentration levels (toxic and subtoxic) of 3,4-methylenedioxypyrovalerone on the metabolic profile of primary mouse hepatocytes, under normothermic and hyperthermic conditions, providing new insights into the mechanism of hepatotoxicity induced by this cathinone derivative as well as the higher risks occurring under hyperthermic conditions [87]. In another study, GC-MS has been used to analyze the metabolomic changes in the rat liver after chlorpyriphos, cadmium, and their mixtures treatment [88]. By this technique, a number of eleven biomarkers have been identified, among them being butanedioic acid, myo-inositol, and urea. Another example is the use of a GC-MS-based metabolomic approach for the investigation of the metabolic mechanism of triptolide-induced reproductive toxicity in order to identify potential novel biomarkers for the early detection of spermatogenesis dysfunction [89].

Several studies are reported for measuring the toxicant levels in biological specimens. A group of four synthetic insecticides from the class of pyrethroids (tefluthrin, bifenthrin, α-cypermethrin, and deltamethrin) were determined collected samples of blood, liver, and cerebellum were analyzed 6 hours after administration with the aid of GC with electron capture detector (GC-ECD) [90]. The results provided information regarding the exposure-dose-effect relation for pyrethroids and were useful for designing pharmacokinetic models for environmentally relevant exposures to pyrethroid mixtures. Polychlorinated biphenyls (PCBs) are another class of toxicants that can be determined by GC. Their concentrations (at levels of pg/g), for example, in maternal blood during pregnancy for 169 participants, and the associations between prenatal exposure to various PCBs and the gene methylation levels were evaluated in infants by GC coupled to high-resolution mass spectrometry (GC-HRMS), after extraction with organic solvent [91].

An important class of applications refers to the investigations of specific pharmaceuticals used for reducing the effect of certain toxicants. For example, sodium salicylate has been shown to be a promising antidote for the treatment of paraquat (N,N′-dimethyl-4,4′-biphenyl dichloride) poisonings [92]. Besides the modulation of the pro-oxidant and pro-inflammatory pathways and anti-thrombogenic properties of sodium salicylate, this study was focused on proving the possibility that a direct chemical reaction may take place between sodium salicylate and paraquat as a result of charge-transfer complexes, whose stoichiometry was established from GC-MS experiments. The possibility of formation of specific adducts in biological matrices can be made using a variety of techniques including GC-MS. An example is the investigation of detoxification of severe nerve agents like cyclosarin, sarin, tabun, and VX (ethyl N-2-diisopropylaminoethyl methylphosphonothiolate) using β-cylodextrin derivatized with iodosobenzoic acid (CD-IBA) [93]. The biochemical assay was based on GC-MS determinations of the nerve agent concentrations in the extracts of chloroform obtained from biomatrix, using propyl-N,N-dimethylphosphoramidocyanidate as internal standard [93]. The possibility of the identification of toxic effects of approved drugs by using GC for their measure in biological fluids is another example of the utilization of this technique. Toxicological analysis by means of GC-MS and GC-MS-TOF for the determination of propofol in the blood and urine were used in real situations of suspected acute and lethal intoxication caused by this pharmaceutical compound [94]. Other examples are the designer drugs: α-pyrrolidinovalerophenone and its metabolites in urine and blood in an acute poisoning case [95, 96], zolpidem in postmortem specimens in a voluntary intoxication [97], antiepileptic drugs (pentobarbital, phenobarbital, and carbamazepine), and antipsychotic drugs (chlorpromazine and thioridazine) in blood samples [98], or drugs and pesticides in postmortem blood [99].

Another toxicological example is the GC determination of two β-carbolines alkaloids (harmine and harmaline), as well as the potent hallucinogen N,N-dimethyltryptamine as the main active components in ayahuasca, which is a hallucinogenic beverage used in religious rituals in South America. In this particular case, sweat was the biological matrix selected for the investigation of these species found in ayahuasca, because this can be collected by a simple and non-invasive procedure, subjected to solid-phase extraction (SPE), and followed by GC-MS analysis [100]. In general, sweat analysis has become a very useful tool in toxicology for monitoring the therapeutic drugs and drugs of abuse [101, 102].

8. Perspectives

Gas chromatography with MS or other types of detection is a mature analytical technique with broad applications in various fields, including bioanalysis. In a few examples of applications of GC in bioanalysis presented in this chapter one may conclude that its performances (separation capacity, detection limits, and complex information provided by MS) are exceeded by other related separation techniques, such as LC-MS. However, GC-MS potential in bioanalysis can be very much enhanced by the versatile sample preparation that makes biocompounds amenable to GC analysis. Generally, the new applications of GC depend on both methodology and instrumental developments and improvements. Advances in sample preparation coupled with on-line GC, high-resolution MS, bidimensional and comprehensive GC, and chemometrics are only a few directions of developing this technique and improving its analytical performances [103, 104]. GC and GC-MS perspectives in bioanalysis have been extended for different types of applications, such as for example in the case of omics-based domains [105, 106].

Author details

Victor David[1*] and Serban C. Moldoveanu[2]

1 Faculty of Chemistry, Department of Analytical Chemistry, University of Bucharest, Bucharest, Romania

2 R.J. Reynolds Tobacco Co., Winston-Salem, NC, USA

*Address all correspondence to: victor.david@chimie.unibuc.ro

References

[1] Poole C. Gas Chromatography. 2nd ed. Amsterdam: Elsevier; 2021

[2] Kocova Vlckova H, Pilarova V, Svobodova P, Plisek J, Svec F, Novakova L. Current state of bioanalytical chromatography in clinical analysis. The Analyst. 2018;**143**:1305-1325

[3] de Aquiar PN, Hantao LW. The role of gas chromatography in bioanalysis. In: Kubota LT, da Silva JA, Sena MM, Alves WA, editors. Tools and Trends in Bioanalytical Chemistry. New York: Springer; 2022

[4] Kazakevich Y, Lobruto R, editors. HPLC for Pharmaceutical Scientists. Hoboken: Wiley-Interscience; 2007

[5] Medvedovici A, Bacalum E, David V. Sample preparation for large scale bioanalytical studies based on liquid chromatographic technique. Biomedical Chromatography. 2018;**32**(1):e4137

[6] Sandra P, David F. Developments in sample preparation for capillary GC analysis. In: Handley AJ, Adlard ER, editors. Gas Chromatographic Techniques and Applications. Sheffield: Academic Press; 2001. pp. 1-50

[7] Moldoveanu SC, David V. Derivatization methods in GC and GC/MS. In: Kusch P, editor. Gas Chromatography: Derivatization, Sample Preparation, Application. London: IntechOpen; 2018

[8] Moldoveanu SC, David V. Modern Sample Preparation for Chromatography. 2nd ed. Amsterdam: Elsevier; 2021

[9] Orata F. Derivatization reactions and reagents for gas chromatography analysis. In: Advanced Gas Chromatography—Progress in Agricultural, Biomedical and Industrial Applications. London: IntechOpen; 2012

[10] Zenkevich IG. Acids: Derivatization for GC analysis. In: Cazes J, editor. Encyclopedia of Chromatography. 3rd ed. Abingdon-on-Thames: Taylor & Francis; 2009

[11] Bloemen HJT, Burn J. Chemistry and Analysis of Volatile Organic Compounds in the Environment. New York: Springer; 1995

[12] Wilde MJ, Cordell RL. Breath analysis by two-dimensional gas chromatography with dual flame ionisation and mass spectrometric detection—Method optimisation and integration within a large-scale clinical study. Journal of Chromatography. A. 2019;**1594**:160-172

[13] Gallager M, Wysocki CJ, Leyden JJ, Spielman AI, Sun X, Pretti G. Analyses of volatile organic compounds from human skin. British Journal of Dermatology. 2008;**159**:780-791

[14] Grabowska-Polanowska B, Faber J, Skowron M, Miarka P, Pietrzycka M, Sliwka I, et al. Detection of potential chronic kidney disease markers in breath using gas chromatography with mass-spectral detection coupled with thermal desorption method. Journal of Chromatography. A. 2013;**1301**:179-189

[15] Probert CSJ, Ahmed I, Khalid T, Johnson E, Smith S, Ratcliffe N. Volatile organic compounds as diagnostic biomarkers in gastrointestinal and liver diseases. Journal of Gastrointestinal and Liver Diseases. 2009;**18**:337-343

[16] Yuan Z-C, Zhang Y, Cai S-H, Chen W, Hu B. Solid phase

microextraction for human breath analysis of environmental and occupational exposures: A review. Advances in Sample Preparation. 2022;**2022**:100023

[17] Di Gilio A, Palmisani J, Ventrella G, Facchini L, Catino A, Varesano N, et al. Breath analysis: Comparison among methodological approaches for breath sampling. Molecules. 2020;**25**(24):5823

[18] Marc M, Zabiegala B, Namiesnik J. Miniaturized passive emission chambers for in situ measurement of emissions of volatile organic compounds. Critical Reviews in Analytical Chemistry. 2013;**43**:55-61

[19] Zeinali S, Pourkar M, Khan K, Patel D, Pawliszyn J. On-site microextraction technologies for the comprehensive investigation of breath composition in lung cancer patients. Green Analytical Chemistry. 2022;**2**:100018

[20] Zeinali S, Pawliszyn J. Protocol for a needle-trap device coupled to GC for the analysis of volatile and semi-volatile compounds in solid and liquid samples. Green Analytical Chemistry. 2022;**2**:100015

[21] Ahmed WM, Lawal O, Nijsen TM, Goodacre R, Fowler SJ. Exhaled volatile organic compounds of infection: A systematic review. ACS Infectious Diseases. 2017;**3**(10):695-710

[22] Ghimenti S, Lomonaco T, Bellagambi FG, Tabucchi S, Onor M, Trivella MG, et al. Comparison of sampling bags for the analysis of volatile organic compounds in breath. Journal of Breath Research. 2015;**9**(4):047110

[23] Song G, Qin T, Liu H, Xu G-B, Pan Y-Y, Xiong F-X, et al. Quantitative breath analysis of volatile organic compounds of lung cancer patients. Lung Cancer. 2010;**67**(2):227-231

[24] Bousamra M, Schumer E, Li M, Knipp RJ, Nantz MH, van Berkel V, et al. Quantitative analysis of exhaled carbonyl compounds distinguishes benign from malignant pulmonary disease. The Journal of Thoracic and Cardiovascular Surgery. 2014;**148**(3):1074-1081

[25] Louhelainen N, Rytilä P, Obase Y, Mäkelä M, Haahtela T, Kinnula VL, et al. The value of sputum 8-isoprostane in detecting oxidative stress in mild asthma. The Journal of Asthma. 2008;**45**:149-154

[26] Al-Attabi Z, D, Arcy BR, Deeth HC. Volatile sulphur compounds in UHT milk. Critical Reviews in Food Science and Nutrition. 2009;**49**:28-47

[27] Musteata ML, Musteata FM. Analytical methods used in conjunction with solid-phase microextraction: A review of recent bioanalytical applications. Bioanalysis. 2009;**1**:1081-1102

[28] Lett L, George M, Slater R, De Lacy CB, Ratcliffe N, García-Finana M, et al. Investigation of urinary volatile organic compounds as novel diagnostic and surveillance biomarkers of bladder cancer. British Journal of Cancer. 2022;**127**(2):329-336

[29] Garner CE, Smith S, Elviss NC, Humphrey TJ, White P, Ratcliffe NM, et al. Identification of Campylobacter infection in chickens from volatile faecal emissions. Biomarkers. 2008;**13**:413-421

[30] Ferre S, Gonzalez-Ruiz V, Guillarme D, Rudaz S. Analytical strategies for the determination of amino acids: Past, present and future trend. Journal of Chromatography B. 2019;**1132**:121819

[31] Derezinski P, Klupczynska A, Sawicki W, Palka JA, Kokot ZJ. Amino acid profiles of serum and urine in search for prostate cancer biomarkers: A pilot study. International Journal of Medical Sciences. 2017;**14**(1):1-12

[32] Nagao K, Kimura T. Use of plasma-free amino acids as biomarkers for detecting and predicting disease risk. Nutrition Reviews. 2020;**78**:79-85

[33] Galezowska G, Ratajczyk J. Determination of amino acids in human biological fluids by high performance liquid chromatography. Amino Acids. 2021;**53**:993-1009

[34] Yu C, Wang L, Zheng J, Jiang X, Zhang Q, Zhang Y, et al. Nanoconfinement effect based in-fiber extraction and derivatization method for ultrafast analysis of twenty amines in human urine by GC-MS: Application to cancer diagnosis biomarkers' screening. Analytica Chimica Acta. 2022;**1217**:339985

[35] Kaspar H, Dettmer K, Chan Q, Daniels S, Nimkar S, Daviglus ML, et al. Urinary amino acid analysis: A comparison of iTRAQ®–LC–MS/MS, GC–MS, and amino acid analyzer. Journal of Chromatography B. 2009;**877**:1838-1846

[36] Hušek P. Quantitation of amino acids as chloroformates—A return to gas chromatography. In: Molnár-Perl I, editor. Quantitation of Amino Acids and Amines by Chromatography. Amsterdam: Elsevier; 2005

[37] Kaspar H, Dettmer K, Gronwald W, Oefner PJ. Automated GC-MS analysis of free amino acids in biological fluids. Journal of Chromatography B. 2008;**870**(2):222-232

[38] Sobolevsky TG, Revelsky AI, Miller B, Oriedo V, Chernetsova ES, Revelsky IA. Comparison of silylation and esterification/acylation procedures in GC-MS analysis of amino acids. Journal of Separation Science. 2003;**26**(17):1474-1478

[39] Li J, Qi H-Y, Wang Y-B, Su Q, Wu S, Wu L. Hollow fiber-stir bar sorptive extraction and microwave assisted derivatization of amino acids in biological matrices. Journal of Chromatography. A. 2016;**1474**:32-39

[40] Paiva MJ, Costa Menezes H, Pereira Christo P, Ribeiro Resende R, de Lourdes Cardeal Z. An alternative derivatization method for the analysis of amino acids in cerebrospinal fluid by gas chromatography-mass spectrometry. Journal of Chromatography B. 2013;**931**:97-102

[41] Pautova A, Burnakova N, Revelsky A. Metabolic profiling and quantitative analysis of cerebrospinal fluid using gas chromatography-mass spectrometry: Current methods and future perspectives. Molecules. 2021;**26**:3597

[42] Hernandez CC, Destaillats F. Gas chromatographic analysis of lipids. In: Gas Chromatography. 2nd edition, (Editor: Poole C.). Amsterdam: Elsevier; 2021. pp. 683-701

[43] Chiu H-H, Kuo C-H. Gas chromatography-mass spectrometry-based analytical strategies for fatty acid analysis in biological samples. Journal of Food and Drug Analysis. 2020;**28**(1):60-73

[44] Hadi NI, Jamal Q, Iqbal A, Shaikh F, Somroo S, Musharraf SG. Serum metabolomic profiles for breast cancer diagnosis, grading and staging by gas chromatography-mass spectrometry. Scientific Reports. 2017;**7**:1715

[45] Budczies J, Denkert C, Müller BM, Brockmöller SF, Klauschen F, Györffy B,

et al. Remodeling of central metabolism in invasive breast cancer compared to normal breast tissue - a GC-TOFMS based metabolomics study. BMC Genomics. 2012;**13**:334

[46] Ruiz-Matute AI, Hernandez-Hernandez O, Rodriguez-Sanchez S, Sanz ML, Martinez-Castro I. Derivatization of carbohydrates for GC and GC-MS analyses. Journal of Chromatography B. 2011;**879**:1226-1240

[47] Kermack AJ, Wellstead SJ, Fisk HL, Cheong Y, Houghton FD, Macklon NS, et al. The fatty acid composition of human follicular fluid is altered by a 6-week dietary intervention that includes marine omega-3 fatty acids. Lipids. 2020;**56**:201-209

[48] Ciucanu CI, Vlad DC, Ciucanu I, Dumitrascu V. Selective and fast methylation of free fatty acids directly in plasma for their individual analysis by gas chromatography- mass spectrometry. Journal of Chromatography. A. 1624;**2020**:461259

[49] Cisbani G, Koppel A, Metherel AH, Smith ME, Aji KN, Andreazza AC, et al. Serum lipid analysis and isotopic enrichment is suggestive of greater lipogenesis in young long-term cannabis users: A secondary analysis of a case–control study. Lipids. 2022;**57**:125-140

[50] Chen S. Lipidomics: Sea Food, Marine Based Dietary Supplement, Fruit and Seed. Kerala: Transworld Research Network; 2012

[51] Moldoveanu SC, Chang Y. Dual analysis of triglycerides from certain common lipids and seed extracts. Journal of Agricultural and Food Chemistry. 2011;**59**:2137-2147

[52] Moldoveanu SC, David V. Sample Preparation in Chromatography. Amsterdam: Elsevier; 2002

[53] MeyerM, MonteroL, MeckelmannSW, Schmitz OJ. Comparative study for analysis of carbohydrates in biological samples. Analytical and Bioanalytical Chemistry. 2022;**414**:2117-2130

[54] Ciucanu I, Kerek F. A simple and rapid method for the permethylation of carbohydrates. Carbohydrate Research. 1984;**131**(2):209-217

[55] Ciucanu I. Per-O-methylation reaction for structural analysis of carbohydrates by mass spectrometry. Analytica Chimica Acta. 2006;**576**:147-155

[56] Ciucanu I, Pilat L,Ciucanu CI, Şişu E. Determination of neutral monosaccharides as per-O-methylated derivatives directly from a drop of whole blood by gas chromatography-mass spectrometry. Analytical Chemistry. 2015;**87**:10856-10861

[57] Grabarics M, Lettow M, Carla Kirschbaum C, Greis K, Manz C, Pagel K. Mass spectrometry-based techniques to elucidate the sugar code. Chemical Reviews. 2022;**122**:7840-7908

[58] Delafield DG, Li L. Recent advances in analytical approaches for glycan and glycopeptide quantitation. Molecular & Cellular Proteomics. 2021;**20**:100054

[59] Pujic I, Perreault H. Recent advancements in glycoproteomic studies: Glycopeptide enrichment and derivatization, characterization of glycosylation in SARS CoV2, and interacting glycoproteins. Mass Spectrometry Reviews. 2022;**41**(3):488-507

[60] Bowden JA, Ford DA. An examination of pentafluorobenzoyl derivatization strategies for the analysis of fatty alcohols using gas chromatography/electron capture

negative ion chemical ionization–mass spectrometry. Journal of Chromatography B. 2011;**879**:1375-1383

[61] Available from: https://www.fda.gov/regulatory-information/search-fda-guidance-documents/m10-bioanalytical-method-validation [Accessed: May 6, 2022]

[62] Bressolle F, Bromet-Petit M, Audran M. Validation of liquid chromatographic and gas chromatographic methods, application to pharmacokinetics. Journal of Chromatography. 1996;**686**:3-10

[63] Segura J, Ventura R, Jurado C. Derivatization procedures for gas chromatographic-mass spectrometric determination of xenobiotics in biological samples, with special attention to drugs of abuse and doping agents. Journal of Chromatography. B, Biomedical Sciences and Applications. 1998;**713**(1):61-90

[64] Salerno TMG, Donato P, Frison G, Zamengo L, Mondello L. Gas chromatography-fourier transform infrared spectroscopy for unambiguous determination of illicit drugs: A proof of concept. Frontiers in Chemistry. 2022;**8**:624

[65] Manousi N, Samanidou VF. Applications of gas chromatography for the analysis of tricyclic antidepressants in biological matrices. Separations. 2019;**6**:24

[66] Brettell TA, Lum BJ. Analysis of drugs of abuse by gas chromatography-mass spectrometry (GC-MS). In: Musah R, editor. Analysis of Drugs of Abuse: Methods in Molecular Biology. Vol. 1810. New York: Humana Press; 2018

[67] Valdez CA. Gas chromatography-mass spectrometry analysis of synthetic opioids belonging to the fentanyl class: A review. Critical Reviews in Analytical Chemistry. 2022. DOI: 10.1080/10408347,2021,1927668

[68] Alves EA, Agonia AS, Cravo SM, Afonso CM, Pereira Netto AD, de Lourdes BM, et al. GC-MS method for the analysis of thirteen opioids, cocaine and cocaethylene in whole blood based on a modified Quechers extraction. Current Pharmaceutical Analysis. 2017;**13**:215-223

[69] Verma P, Bajaj A, Tripathi RM, Shukla SK, Nagpal S. Development and validation of the method for the detection of glimepiride via derivatization employing N-methyl-N-(trimethylsilyl) trifluoroacetamide using gas chromatography-mass spectrometry. Egyptian Journal of Forensic Sciences. 2021;**11**:4

[70] Colucci AP, Aventaggiato L, Centrone M, Gagliano-Candela R. Validation of an extraction and gas chromatography—Mass spectrometry quantification method for cocaine, methadone, and morphine in postmortem adipose tissue. Journal of Analytical Toxicology. 2010;**34**:342-346

[71] Cabarcos-Fernandez P, Alvarez-Freire I, Tabernero-Duque MJ, Bermejo-Barrera AM. Quantitative determination of clozapine in plasma using an environmentally friendly technique. Microchemical Journal. 2022;**180**:107612

[72] Rodrigues A, Yegles M, Van Elsue N, Schneider S. Determination of cannabinoids in hair of CBD rich extracts consumers using gas chromatography with tandem mass spectrometry (GC/MS-MS). Forensic Science International. 2018;**292**:163-166

[73] Shibasaki H, Nakayama H, Furuta T, Kasuya Y, Tsuchiya M, Soejima A, et al.

Simultaneous determination of prednisolone, prednisone, cortisol, and cortisone in plasma by GC-MS: Estimating unbound prednisolone concentration in patients with nephrotic syndrome during oral prednisolone therapy. Journal of Chromatography B. 2008;**870**:164-169

[74] Saraji M, Boroujeni MK. Analysis of narcotic drugs in biological samples using hollow fiber liquid-phase microextraction and gas chromatography with nitrogen phosphorus detection. Microchimica Acta. 2011;**174**:159-166

[75] Alsenedi KA, Morrison C. Determination of amphetamine-type stimulants (ATSs) and synthetic cathinones in urine using solid phase micro-extraction fibre tips and gas chromatography-mass spectrometry. Analytical Methods. 2018;**10**(12):1431-1440

[76] Molnar B, Fodor B, Csampai A, Hidvegi E, Molnár-Perl I. Structure-related new approach in the gas chromatography/mass spectrometry analysis of cathinone type synthetic drugs. Journal of Chromatography. A. 2016;**1477**:70-75

[77] Zoerner AA, Heusser K, Gutzki FM, Mitschke A, Tank J, Stichtenoth DO, et al. Unique pentafluorobenzylation and collision-induced dissociation for specific and accurate GC-MS/MS quantification of the catecholamine metabolite 3,4-dihydroxyphenylglycol (DHPG) in human urine. Journal of Chromatography B. 2011;**879**:1444-1456

[78] Yilmaz B, Arslan S. Determination of carvedilol in human plasma by gas chromatography - mass spectrometry method. Journal of Chromatographic Science. 2011;**49**:35-39

[79] Shah I, Barker J, Naughton DP, Barton SJ, Ashraf SS. Determination of diclofenac concentrations in human plasma using a sensitive gas chromatography mass spectrometry method. Chemistry Central Journal. 2016;**10**:52

[80] Baskal S, Bollenbach A, Henzi B, Hafner P, Fischer D, Tsikas D. Stable-isotope dilution GC-MS measurement of metformin in human serum and urine after derivatization with pentafluoropropionic anhydride and its application in Becker muscular dystrophy patients administered with metformin, L-citrulline, or their combination. Molecules. 2022;**27**:3850

[81] Silvestro L, Tarcomnicu I, Rizea SS. Matrix effects in mass spectrometry combined with separation methods—Comparison HPLC, GC and discussion on methods to control these effects. In: Coelho AV, Matos Ferraz Franco C, editors. Tandem Mass Spectrometry. London: IntechOpen; 2013

[82] Tsuchiyama T, Miki Katsuhara M, Masahiro NM. Compensation of matrix effects in gas chromatography-mass spectrometry analysis of pesticides using a combination of matrix matching and multiple isotopically labeled internal standards. Journal of Chromatography. A. 2017;**1524**:233-245

[83] Kwon H, Anastassiades M, Dork D, Hong SM, Moon BC. Compensation for matrix effects in GC analysis of pesticides by using cucumber extract. Analytical and Bioanalytical Chemistry. 2018;**410**(22):5481-5489

[84] Singh Z. Forensic toxicology: Biological sampling and use of different analytical techniques. Forensic Research & Criminology International Journal. 2017;**4**(4):117-120

[85] Favretto D, Montisci M, Snenghi R. Forensic toxicology. In: Andreazza AC,

Scola G, editors. Toxicology Studies. London: IntechOpen; 2015

[86] Buah-Kwofie A, Humphries MS. Validation of a modified QuEChERS method for the analysis of organochlorine pesticides in fatty biological tissues using two-dimensional gas chromatography. Journal of Chromatography B. 2019;**1105**:85-92

[87] Araujo AM, Bastos M, Carvalho F, Guedes de Pinho P, Carvalho M. Effect of temperature on 3,4-methylenedioxypyrovalerone (MDPV)-induced metabolome disruption in primary mouse hepatic cells. Toxicology. 2020;**441**:152503

[88] Xu M-Y, Wang P, Sun Y-J. Metabolomic analysis for combined hepatotoxicity of chlorpyrifos and cadmium in rats. Toxicology. 2017;**384**:50-58

[89] Ma B, Qi H, Li J, Xu H, Chi B, Zhu J, et al. Triptolide disrupts fatty acids and peroxisome proliferator-activated receptor (PPAR) levels in male mice testes followed by testicular injury: A GC-MS based metabolomics study. Toxicology. 2015;**336**:84-95

[90] Mosquera Ortega ME, Romero DM, Pato AM, Sosa-Holt CS, Ridolfi A, Villaamil Lepori E, et al. Relationship between exposure, body burden and target tissue concentration after oral administration of a low-dose mixture of pyrethroid insecticides in young adult rats. Toxicology. 2018;**409**:53-62

[91] Kobayashi S, Sata F, Miyashita C, Miura R, Azumi K, Kobayashi S, et al. Gender-specific association of exposure to non-dioxin-like polychlorinated biphenyls during pregnancy with methylation levels of H19 and long interspersed nuclear element-1 in cord blood in the Hokkaido study. Toxicology. 2017;**390**:135-145

[92] Dinis-Oliveira RJ. Reactivity of paraquat with sodium salicylate: Formation of stable complexes. Toxicology. 2008;**249**:130-139

[93] Wille T, Tenberken O, Reiter G, Müller S, Le Provost R, Lafont O, et al. Detoxification of nerve agents by a substituted β-cyclodextrin: Application of a modified biological assay. Toxicology. 2009;**265**:96-100

[94] Procaccianti P, Fare F, Argo A, Casagni E, Arnoldi S, Facheris S, et al. Determination of Propofol by GC/MS and fast GC/MS-TOF in two cases of poisoning. Journal of Analytical Toxicology. 2017;**41**:771-776

[95] Grapp M, Sauer C, Vidal C, Müller D. GC–MS analysis of the designer drug α-pyrrolidinovalerophenone and its metabolites in urine and blood in an acute poisoning case. Forensic Science International. 2016;**259**:e14-e19

[96] Grapp M, Kaufmann C, Ebbecke M. Toxicological investigation of forensic cases related to the designer drug 3,4-methylenedioxypyrovalerone (MDPV): Detection, quantification and studies on human metabolism by GC–MS. Forensic Science International. 2016;**273**:1-9

[97] Keller T, Schneider A, Tutsch-Bauer E. GC/MS determination of zolpidem in postmortem specimens in a voluntary intoxication. Forensic Science International. 1999;**106**(2):103-108

[98] Bravo F, Ulloa R, Venegas K, Contzen M, Rios D, Briceno M, et al. Validation of a method to detect central nervous system acting drugs in blood by GC/MS; application in cases of death caused by phenothiazines in the north of

Chile. Journal of the Chilean Chemical Society. 2014;**59**:2602-2605

[99] Junior EF, Dutra CE. Simultaneous determination of drugs and pesticides in postmortem blood using dispersive solid-phase extraction and large volume injection-programmed temperature vaporization-gas chromatography–mass spectrometry. Forensic Science International. 2018;**290**:318-326

[100] Tavares L, Monedeiro F, Bordin DM, De Martinis BS. Investigation of ayahuasca β-carboline alkaloids and tryptamine in sweat samples from religious community participants by GC-MS. Journal of Analytical Toxicology. 2020;**44**:601-609

[101] de Giovanni N, Fucci N. The current status of sweat testing for drugs of abuse: A review. Current Medicinal Chemistry. 2013;**20**(4):545-561

[102] Jadoon S, Karim S, Akram MR, Khan AK, Zia MA, Siddiqi AR, et al. Recent developments in sweat analysis and its applications. International Journal of Analytical Chemistry. 2015;**2015**:164974

[103] Prebihalo SE, Berrier KL, Freye CE, Bahaghighat HD, Moore NR, Pinkerton DK, et al. Multidimensional gas chromatography: Advances in instrumentation, chemometrics, and applications. Analytical Chemistry. 2018;**90**(1):505-532

[104] Gruber B, David F, Sandra P. Capillary gas chromatography-mass spectrometry: Current trends and perspectives. TrAC, Trends in Analytical Chemistry. 2020;**124**:115475

[105] Koek MM, Jellema RH, van der Greef J, Tas AC, Hankemeier T. Quantitative metabolomics based on gas chromatography mass spectrometry: Status and perspectives. Metabolomics. 2011;**7**:307-328

[106] Sandra K, Sandra P. Lipidomics from an analytical perspective. Current Opinion in Chemical Biology. 2013;**17**(5):847-853

Chapter 6

Designs for Screening Experiments with Quantitative Factors

Nam-Ky Nguyen, Stella Stylianou, Tung-Dinh Pham and Mai Phuong Vuong

Abstract

Most screening experiments in chemometrics and science are quantitative, i.e. continuous factors. These factors should be 3-level and the designs for these experiments should also be 3-level. However, popular designs for screening experiments are still Plackett-Burman designs (PBDs) and 2-level fractional factorial designs (FFDs) such as resolution III and resolution IV FFDs. This chapter introduces the conference matrices as an alternative to PBDs and resolution III FFDs and definitive screening designs, a conference matrix-based class of designs, as an alternative to resolution IV FFDs. A table of conference matrices of up to order 32 and examples are also provided for illustration.

Keywords: conference matrices, definitive screening designs, fractional factorial designs, Plackett-Burman designs, response surface designs, screening designs

1. Introduction

Screening experiments are used at the initial stage of experimentation and aim at identifying the dominant main effects out of a large set of potentially active factors. The benefit of the screening approach is the use of a cost-effective design and process to separate the influential variables from the non-influential ones. By using the active effects that have been identified by the screening process, the research can run additional follow-up experiments to fit higher-order effects and build a better and more complex model. Screening is traditionally performed by applying a linear model using a 2-level FFD. When the screening process involves quantitative factors, other designs with better properties are also available in the recent literature.

Consider a 2^{15-11} experiment conducted by Poorna & Kulkarnin [1] (hereafter abbreviated as PK) to investigate 15 2-level factors, which might affect inulinase production. These factors come from four carbon sources: **A** Inulin (%), **B** Fructose (%), **C** Glucose (%), **D** Sucrose (%); four organic nitrogen sources: **E** Corn steep liquor (%), **F** Peptone (%), **G** Urea (%), **H** Yeast extract (%); four inorganic nitrogen sources: **J** Corn steep liquor (%), **K** Peptone (%), **L** Urea (%), **M** Yeast extract (%); and three other parameters: **N** Trace element solutions (mL), **O** Inoculum level (10^6 spores/mL), **P** (pH). The two responses are inulinase activity (units/mL) at

60 hours and dry weight biomass (mg/mL). This experiment is also summarized in Example 6.4 of [2].

The design for the PK experiment is a 2_{III}^{15-11} (a resolution III FFD for 15 factors in 16 runs) given in **Table 1(a)**. The 11 design generators for this experiment are: $\mathbf{E} = -\mathbf{ABCD}, \mathbf{F} = \mathbf{BCD}, \mathbf{G} = \mathbf{ABC}$, $\mathbf{H} = -\mathbf{CD}, \mathbf{I} = -\mathbf{BD}, \mathbf{J} = \mathbf{ABD}, \mathbf{K} = \mathbf{ACD}$, $\mathbf{L} = -\mathbf{AC}, \mathbf{M} = -\mathbf{AD}, \mathbf{N} = -\mathbf{AB}$, and $\mathbf{O} = -\mathbf{BC}$. For a resolution III design, no main effects (MEs) are aliased with any other MEs, but MEs are aliased with 2-factor interactions (2FIs). For this design, each ME is aliased with seven 2FIs. For example $-\mathbf{A} = \mathbf{BN} = \mathbf{CL} = \mathbf{DM} = \mathbf{EF} = \mathbf{GO} = \mathbf{HK} = \mathbf{IJ}$ and $-\mathbf{D} = \mathbf{AM} = \mathbf{BI} = \mathbf{CH} = \mathbf{EG} = \mathbf{FO} = \mathbf{JN} = \mathbf{KL}$. An alternative design with similar number of runs, whose MEs are pairwise orthogonal and are not fully aliased with 2FIs, will be presented in this chapter.

Let us examine another experiment on the human blood formation that originates in hematopoietic stem cells (HSC) and hematopoietic progenitor cells (HPC). Due to the complexity of clinical use of cells in serum medium, Yao et.al. [3] set up a 2^{8-4} experiment to screen out the most important serum substitutes affecting the growth of HSC and HPC among eight kinds of compounds: **A** Albumax I (10g/l), **B** BSA (10g/l), **C** TF(0.4g/l), **D** Glutamine (2mM), **E** HC (1mg/l), **F** Peptone (1g/l), **G** 2-ME

(a)

A	B	C	D	E	F	G	H	I	J	K	L	M	N	O
1	1	1	1	-1	1	1	-1	-1	1	1	-1	-1	-1	-1
1	1	1	-1	1	-1	1	1	1	-1	-1	-1	1	-1	-1
-1	1	1	1	1	1	-1	-1	-1	-1	-1	1	1	1	-1
1	-1	1	-1	-1	1	-1	1	-1	1	-1	-1	1	1	1
-1	1	-1	1	-1	-1	1	1	-1	-1	1	-1	1	1	1
1	-1	1	1	1	-1	-1	-1	1	-1	1	-1	-1	1	1
1	1	-1	-1	-1	1	-1	-1	1	-1	1	1	1	-1	1
-1	1	1	-1	-1	-1	-1	1	1	1	1	1	-1	1	-1
-1	-1	1	1	-1	-1	1	-1	1	1	-1	1	1	-1	1
1	-1	-1	-1	1	-1	1	-1	-1	1	1	1	1	1	-1
-1	1	-1	-1	1	1	1	-1	1	1	-1	-1	-1	1	1
-1	-1	1	-1	1	1	1	1	-1	-1	1	1	-1	-1	1
-1	-1	-1	1	1	1	-1	1	1	1	1	-1	1	-1	-1
1	-1	-1	1	-1	1	1	1	1	-1	-1	1	-1	1	-1
1	1	-1	1	1	-1	-1	1	-1	1	-1	1	-1	-1	1
-1	-1	-1	-1	-1	-1	-1	-1	-1	-1	-1	-1	-1	-1	-1

(b)

A	B	C	D	E	F	G	H
-1	-1	-1	-1	-1	-1	-1	-1
1	-1	-1	-1	-1	1	1	1
-1	1	-1	-1	1	-1	1	1
1	1	-1	-1	1	1	-1	-1
-1	-1	1	-1	1	1	1	-1
1	-1	1	-1	1	-1	-1	1
-1	1	1	-1	-1	1	-1	1
1	1	1	-1	-1	-1	1	-1
-1	-1	-1	1	1	1	-1	1
1	-1	-1	1	1	-1	1	-1
-1	1	-1	1	-1	1	1	-1
1	1	-1	1	-1	-1	-1	1
-1	-1	1	1	-1	-1	1	1
1	-1	1	1	-1	1	-1	-1
-1	1	1	1	1	-1	-1	-1
1	1	1	1	1	1	1	1

Table 1.
(a) Poorna & Kulkarni's 2^{15-11} *experiment, (b) Yao et al.'s* 2^{8-4} *experiment.*

(55μM) and **H** Insulin (10μg/ml) for hematopoietic ex vivo expansion culture. Among them, the first three factors are non-hormonal proteins, Insulin and HC are both hormonal proteins, 2-ME is an antioxidant molecule, and Glutamine is an amino acid.

The design for the experiment in the second example is a 2_{IV}^{8-4} (a resolution IV FFD for eight factors in 16 runs) given in **Table 1(b)** where level -1 means no addition, and 1 indicates a specified concentration of the compound. The four design generators for this experiment are: $\mathbf{E} = \mathbf{BCD}$, $\mathbf{F} = \mathbf{ACD}$, $\mathbf{G} = \mathbf{ABC}$, $\mathbf{H} = \mathbf{ABD}$. For a resolution IV design, no MEs are aliased with any other ME or 2FIs, but 2FIs are aliased with other 2FIs. For this design, each of 2FIs is aliased with three other 2FIs: $\mathbf{AB} = \mathbf{EF} = \mathbf{CG} = \mathbf{DH}, \mathbf{AC} = \mathbf{DF} = \mathbf{BG} = \mathbf{EH}, \mathbf{AD} = \mathbf{CF} = \mathbf{EF} = \mathbf{BH}, \mathbf{AE} = \mathbf{BF} = \mathbf{DG} = \mathbf{CH}$, $\mathbf{AF} = \mathbf{CD} = \mathbf{BE} = \mathbf{GH}, \mathbf{AG} = \mathbf{BC} = \mathbf{DE} = \mathbf{FH}$, and $\mathbf{AH} = \mathbf{BD} = \mathbf{CE} = \mathbf{FG}$. Again, in this chapter we will present an alternative design, whose MEs are orthogonal to other MEs and 2FIs and whose 2FIs are not fully aliased with other 2FIs.

Motivated by the above examples, this chapter introduces the use of conference matrices and conference matrix-based designs, including the popular definitive screening designs (DSDs), as an alternative to Plackett-Burman designs or PBDs [4] and resolution III and IV FFDs.

2. Conference matrices and its use

A *conference matrix* C of order m is an $m \times m$ $(0, \pm 1)$-matrix with zero diagonal satisfying the condition $\mathbf{CC}' = (m-1)\,\mathbf{I}$, where $\mathbf{I}$ is the identity matrix. A conference matrix is said to be *normalized* if all entries in its first row and first column are 1 (except the (1,1) entry, which is 0). Removing the first row and the first column of a normalized conference matrix yields its *core*. A conference matrix is said to be *skew-symmetric* if $\mathbf{C} = -\mathbf{C}'$. It is conjectured that a conference matrix $\mathbf{C}$ exists for all $m \equiv 2$ (mod 4) as long as $m-1$ is a sum of two squares. Examples of non-existent conference matrices are the ones of size 2, 34 and 58. More information about the conference matrices can be found in Section 6.1 of [5].

A large number of conference matrices can be constructed by the single cyclic generators. **Table 2** displays two generating vectors for the conference matrices with $m \leq 32$. To generate the conference matrix for $m = 8$, for example, we use the generating vector $(0 + + - + - -)$ to generate its core and then augment it with 0 in the (1,1) entry, -1 in the remaining entries of the first row, and +1 in the remaining entries in the first column. Note that if we replace the first element of this generating vector, i.e. 0 by +1, we have the generating vector for the PBD with eight runs.

Table 3 displays the conference matrices for $m = 6, 8, 10, 12, 14, 16, 26$ and 28. Note that, the conference matrices for $m = 10, 16, 26$ and 28 cannot be generated by the cyclic generators. Unlike conference matrices of order $m = 10, 16, 26$ and 28 in [6], the numbers of 1's and -1's in each column from 2 to m differ by only one. A conference matrix of order $2m$ can be constructed from a conference matrix of order m by the following equation:

$$\begin{pmatrix} \mathbf{C} & -(\mathbf{C}+\mathbf{I})' \\ \mathbf{C}+\mathbf{I} & \mathbf{C}' \end{pmatrix} \tag{1}$$

Here, $\mathbf{C}$ in (1) is a conference matrix of order m, and $\mathbf{I}$ is the identity matrix. We use (1) to construct the conference matrix of order 16 in **Table 3** from the conference matrix of order 8 in this table.

m	**Generating vectors**
4	0 − +
6	0 − + + −
8†	0 + + − + − −
10	See **Table 3**.
12†	0 + − + + + − − − + −
14	0 + − + + − − − − + + − +
16	See **Table 3**.
18	0 − − + − + + + − − + + + − + − −
20†	0 + − − + + + + − + − + − − − − + + −
22	Not exist.
24†	0 + + + + − + − + + − − + + − − + − + − − − −
26	See **Table 3**.
28	See **Table 3**.
30	0 + − − + + + + − + − − − + − − + − − − + − + + + + − − +
32	0 + + − + + − + + + + − − − + − + − + + + − − − − + − − + − −

†*Obtained from the generating vector for a Plackett-Burman design. The remaining generating vectors are from* ***Table 1*** *of [6].*

Table 2.
Generating vectors for conference matrices with $m \leq 32$.

One of the most popular use of the conference matrices is to construct *definitive screening designs* (DSDs), which are the 3-level designs introduced in [7] for studying quantitative factors. The design matrix **D** for a DSD can be written as:

$$\begin{pmatrix} \mathbf{C} \\ \mathbf{0} \\ -\mathbf{C} \end{pmatrix} \qquad (2)$$

where **C** is a constituent $m \times m$ $(0, \pm 1)$-matrix with zero diagonal, $-C$ is the foldover fraction of **C**, and **0** is a row vector of 0's. Note that **0** can contain more than one row vector of 0's.

The model for a 3-level screening design such as a DSD is:

$$\mathbf{y} = \mathbf{X}\boldsymbol{\beta} + \varepsilon \qquad (3)$$

where **y** is the response vector; **X** is the model matrix of size $n \times p$ with $p = 1 + 2m + \binom{m}{2}$; $\boldsymbol{\beta}$'s is the vector of parameters to be estimated; $\boldsymbol{\varepsilon}$ is the error vector with components assumed to be independent and identically distributed (iid) $N(0,\sigma^2)$. Let $d_{ui},(u = 1, \dots, n; i = 1, \dots, m)$ be the entry in the uth row and ith column of the design matrix **D**. The uth row of **X** can be written as $\left(1, d_{u1}, \dots, d_{um}, d_{u1}^2, \dots, d_{um}^2, d_{u1}d_{u2}, \dots, d_{u(m-1)}d_{um}\right)$. The terms in each vector correspond to the intercept, MEs, quadratic effects (QEs) and 2FIs.

DSDs have the following desirable properties:

i. The design is mean orthogonal;

ii. The number of runs is $n = 2m + 1$, i.e. saturated for estimating the intercept, m MEs and m QEs;

m=6	m=8	m=10	m=12	m = 14
0-----	0-------	0---------	0-----------	0-------------
+0-++-	+0++-+--	+0----++++	+0+-+++---+-	+0+-++----++-+
+-0-++	+-0++-+-	+-0+-+--++	+-0+-+++---+	++0+-++----++-
++-0-+	+--0++-+	+-+0+-+--+	++-0+-+++---	+-+0+-++----++
+++-0-	++--0++-	+--+0+++--	+-+-0+-+++--	++-+0+-++----+
+-++-0	+-+--0++	+-+-+0-++-	+--+-0+-+++-	+++-+0+-++----
	++-+--0+	++-++-0-+-	+---+-0+-+++	+-++-+0+-++---
	+++-+--0	++--++-0-+	++---+-0+-++	+--++-+0+-++--
		+++--++-0-	+++---+-0+-+	+---++-+0+-++-
		++++---+-0	++++---+-0+-	+----++-+0+-++
			+-+++---+-0+	++----++-+0+-+
			++-+++---+-0	+++----++-+0+-
				+-++----++-+0+
				++-++----++-+0

m = 16	m = 26	m = 28
0---------------	0-------------------------	0---------------------------
+0++-+--+-++-+--	+0------------++++++++++++	+0+++++++++++++-------------
+-0++-+-+--++-+-	+-0+--++-++--+++----+-+-++	+-0+++-++--+---+++-+-++-+---
+--0++-++---++-+	+-+0+--++-++--+++----+-+-+	+--0+++-++--+---+++-+-++-+--
++--0++-++---++-	+--+0+--++-++-++++----+-+-	+---0+++-++--+---+++-+-++-+-
+-+--0+++-+---++	+---+0+--++-++-++++----+-+	+----0+++-++--+---+++-+-++-+
++-+--0+++-+---+	+-+--+0+--++-++-++++----+-	+-+---0+++-++--+---+++-+-++-
+++-+--0+++-+---	+-++--+0+--++--+-++++----+	+--+---0+++-++--+---+++-+-++
+-------0+++++++	+--++--+0+--+++-+-++++----	+---+---0+++-+++-+---+++-+-+
++++-+---0--+-++	+-+-++--+0+--+-+-+-++++---	+-+--+---0+++-+++-+---+++-+-
+-+++-+--+0--+-+	+-++-++--+0+----+-+-++++--	+-++--+---0+++--++-+---+++-+
+--+++-+-++0--+-	+--++-++--+0+----+-+-++++-	+--++--+---0++++-++-+---+++-
++--+++---++0--+	+---++-++--+0+----+-+-++++	+-+-++--+---0++-+-++-+---+++
+-+--+++-+-++0--	+-+--++-++--+0+----+-+-+++	+-++-++--+---0++-+-++-+---++
++-+--++--+-++0-	+++++-+-+----+0-++--+--++-	+-+++-++--+---0++-+-++-+---+
+++-+--+---+-++0	++++++-+-+-----0-++--+--++	++-+++-+--+-+--0---+--++-+++
	++-++++-+-+---+-0-++--+--+	++--+++-+--+-+-+0---+--++-++
	++--++++-+-+--++-0-++--+--	++---+++-+--+-+++0---+--++-+
	++---++++-+-+--++-0-++--+-	+++---+++-+--+-+++0---+--++-
	++----++++-+-+--++-0-++--+	++-+---+++-+--+-+++0---+--++
	+++----++++-+-+--++-0-++--	+++-+---+++-+--+-+++0---+--+
	++-+----++++-+-+--++-0-++-	++-+-+---+++-+-++-+++0---+--
	+++-+----++++---+--++-0-++	++--+-+---+++-+-++-+++0---+-
	++-+-+----+++++--+--++-0-+	+++--+-+---+++---++-+++0---+
	+++-+-+----+++++--+--++-0-	++-+--+-+---++++--++-+++0---
	++++-+-+----++-++--+--++-0	+++-+--+-+---++-+--++-+++0--
		++++-+--+-+---+--+--++-+++0-
		+++++-+--+-+------+--++-+++0

Table 3.
Conference matrices for m = *6, 8, 10, 12, 14, 16 and 28.*

iii. If a conference matrix (or some columns of a conference matrix) is used for **C** in (2), the constructed DSD is also orthogonal for MEs [6, 8, 9];

iv. Unlike the resolution III FFDs, the MEs are orthogonal to all 2FIs;

v. Unlike the resolution IV FFDs, 2FIs are not fully aliased with one another;

vi. The number of runs for DSDs are more flexible than the ones for resolution IV FFDs. Unlike the former, the latter should be a $2^k, (k \geq 2)$.

An application of a DSD to decolourization of an azo dye on boron-doped diamond electrodes is given in [10].

3. Discussion

The suggested alternative design for the 2_{III}^{15-11} FFD used in the PK experiment is presented in **Table 4(a)**. This design contains columns 2–16 of the conference matrix of order 16 in **Table 3** and a row of 0's. The 2_{III}^{15-11} FFD has 16 runs, and the suggested alternative has 17 runs. This type of saturated or near-saturated designs is used when the experimental resource is expensive.

To have a complete picture of the aliasing patterns of the PK design and the suggested one, we show the correlation cell plots (CCPs) of the two designs **Figure 1(a)** and **(b)**. These CCPs, proposed by [7], display the magnitude of the correlation (in terms of the absolute values) between m MEs and $\binom{m}{2}$ 2FIs in screening designs. The colour of each cell ranges from white (no correlation) to dark (correlation of 1, which means full aliases). It can be seen that while the cells in CCP for the FFD in **Figure 1(a)** are either white or

(a)

A	B	C	D	E	F	G	H	I	J	K	L	M	N	O
-1	-1	-1	-1	-1	-1	-1	-1	-1	-1	-1	-1	-1	-1	-1
0	1	1	-1	1	-1	-1	1	-1	1	1	-1	1	-1	-1
-1	0	1	1	-1	1	-1	1	-1	-1	1	1	-1	1	-1
-1	-1	0	1	1	-1	1	1	-1	-1	-1	1	1	-1	1
1	-1	-1	0	1	1	-1	1	1	-1	-1	-1	1	1	-1
-1	1	-1	-1	0	1	1	1	-1	1	-1	-1	-1	1	1
1	-1	1	-1	-1	0	1	1	1	-1	1	-1	-1	-1	1
1	1	-1	1	-1	-1	0	1	1	1	-1	1	-1	-1	-1
-1	-1	-1	-1	-1	-1	-1	0	1	1	1	1	1	1	1
1	1	1	-1	1	-1	-1	-1	0	-1	-1	1	-1	1	1
-1	1	1	1	-1	1	-1	-1	1	0	-1	-1	1	-1	1
-1	-1	1	1	1	-1	1	-1	1	1	0	-1	-1	1	-1
1	-1	-1	1	1	1	-1	-1	-1	1	1	0	-1	-1	1
-1	1	-1	-1	1	1	1	-1	1	-1	1	1	0	-1	-1
1	-1	1	-1	-1	1	1	-1	-1	1	-1	1	1	0	-1
1	1	-1	1	-1	-1	1	-1	-1	-1	1	-1	1	1	0
0	0	0	0	0	0	0	0	0	0	0	0	0	0	0

(b)

A	B	C	D	E	F	G	H
0	-1	-1	-1	-1	-1	-1	-1
1	0	1	1	-1	1	-1	-1
1	-1	0	1	1	-1	1	-1
1	-1	-1	0	1	1	-1	1
1	1	-1	-1	0	1	1	-1
1	-1	1	-1	-1	0	1	1
1	1	-1	1	-1	-1	0	1
1	1	1	-1	1	-1	-1	0
0	0	0	0	0	0	0	0
0	1	1	1	1	1	1	1
-1	0	-1	-1	1	-1	1	1
-1	1	0	-1	-1	1	-1	1
-1	1	1	0	-1	-1	1	-1
-1	-1	1	1	0	-1	-1	1
-1	1	-1	1	1	0	-1	-1
-1	-1	1	-1	1	1	0	-1
-1	-1	-1	1	-1	1	1	0

Table 4.
(a) Suggested design for Poorna & Kulkarni's 2^{15-11} *experiment, (b) Suggested design for Yao et al.'s* 2^{8-4} *experiment.*

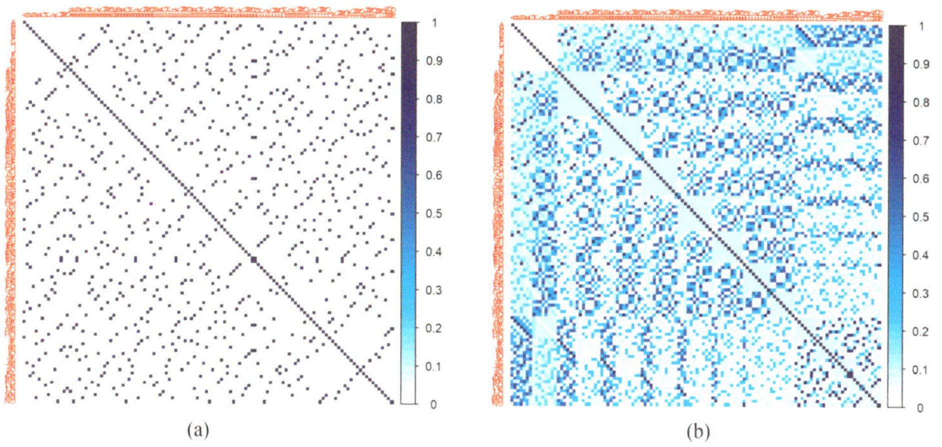

Figure 1.
CCPs for (a) a 2_{III}^{15-11} *FFD used in the Poorna & Kulkarni's experiment and (b) a suggested design for 15 factors in 17 runs in (**Table 4(a)**).*

dark, there are no dark cells in the suggested design in **Figure 1(b)**, meaning none of the MEs (or 2FIs) are fully aliased with the other 2FIs. Altogether, there are 420 dark cells (which represent full aliases) in the upper/lower diagonal portion of the CCP in **Figure 1 (a)**: 105 between the MEs and 2FIs, and 350 among the 2FIs.

Although the first-order D-efficiency of the 2_{III}^{15-11} FFD is higher than that of the suggested design (1 vs. 0.886), and there is a small correlation among the MEs of the suggested design ($|r| = 0.004$), there are at least three reasons for the researchers to choose the latter:

i. Like the resolution III FFD or the PBD, the half fraction of a 3-level DSD might give researchers conclusions similar to the one obtained when the full DSD is used. Consider the data given in **Table 2** in [10] collected from a DSD for 9 factors in 21 runs (one run is a centre run). Analysing the data using the main-effects model, we found two factors **2** and **3** significant at 10% level and two factors **7** and **8** significant at 1% level with the adjusted $R^2 = 0.6775$. Repeating the analysis with the first half fraction of the design (runs with odd order number) plus the centre run, we found three factors **2**, **3**, and **4** significant at 10% level and two factors **7** and **8** significant at 5% level with the adjusted $R^2 = 0.9875$.

ii. The experiment is conducted in stages, and in the first stage, the design is a half fraction of a 3-level DSD and not of the one of a 2-level FFD of resolution IV. In the second stage, the fraction is the foldover of the first half fraction.

iii. The researchers do not wish to use designs with full aliases between MEs and 2FIs and among 2FIs.

The suggested alternative design for the 2_{IV}^{8-4} FFD is the DSD in **Table 4(b)**. This DSD is constructed by Eq. (2) with C being the conference matrix of order 8 in **Table 3**. The CCPs of the 2_{IV}^{8-4} FFD and the DSD are in **Figure 2(a)** and **(b)**, respectively. As expected, both CCPs show that the MEs are orthogonal to the 2FIs. There are 42 dark

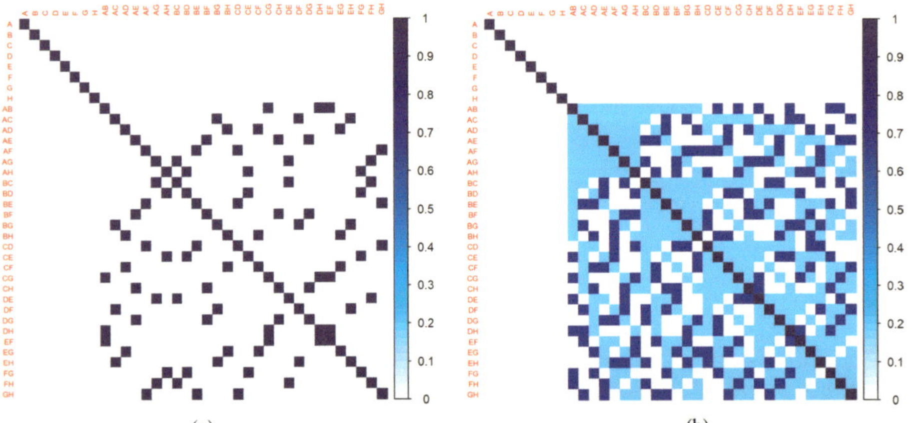

(a) (b)

Figure 2.
*CCPs for (a) a 2_{IV}^{8-4} FFD used in Yao's experiment and (b) a DSD for eight factors in 17 runs (**Table 4(b)**).*

cells (which represent full aliases between the 2FIs) in the upper/lower diagonal portion of the CCP in **Figure 2(a)**. Unlike the CCP for the 2_{IV}^{8-4} FFD in **Figure 1(a)**, the one for the DSD in **Figure 2(b)** shows that none of the 2FIs is fully aliased with the other 2FIs.

The reasons for using a DSD instead of a resolution IV FFD are mentioned in the previous section. Unlike resolution IV FFDs, DSDs can also estimate m QEs (in addition to the intercept and m MEs). Also, QEs are orthogonal to MEs and not fully aliased with 2FIs.

Up to this point, we have been discussing conference matrix-based designs when all factors are quantitative. When there are m_3 3-level factors and m_2 2-level factors, i.e. qualitative or categorical factors, we select $m_3 + m_2$ columns from m columns of a conference matrix and then change 0's to 1's in the last m_2 columns. The uth row of the model matrix $\mathbf{X}$ in (3) is now written as

$$\left(1, d_{u1}, \ldots, d_{u(m_3+m_2)}, d_{u1}^2, \ldots, d_{u(m_3+m_2)}^2, d_{u1}d_{u2}, \ldots, d_{u(m_3+m_2-1)}d_{u(m_3+m_2)}\right).$$

Let C^* be the matrix formed by these columns, the final design is of the form:

$$\begin{pmatrix} \mathbf{C}^* \\ -\mathbf{C}^* \end{pmatrix} \tag{4}$$

This simple method of constructing conference matrix-based designs with mixed-level was mentioned in [11]. Another method of constructing this type of design was discussed in [12]. When there are more 2-level factors than 3-level ones, the readers are encouraged to use the Hadamard matrix-based designs discussed in [13]. Note that when there is a need to block a design into two blocks, one of the 2-level factors can be used as a blocking factor. When there is a need to block a design into three or more blocks or when there is more than one blocking factor, readers are encouraged to refer to [14].

4. Conclusions

This chapter advocates the use of conference matrix-based designs for screening experiments when the factors are quantitative. For these experiments, a number of quantitative factors have to be studied, but only a few of them is expected to be important. In the past, popular designs for this type of experiment are PBDs and 2-level regular FFDs of resolution III and IV. Conference matrix-based designs, unlike the mentioned popular designs, cannot be analysed by hand-calculators. Nowadays, this is not an issue as most data analysis can be done on computers.

Data from experiments using conference matrix-based designs discussed in this paper can be analysed by more advanced statistical methods such as subset or step-wise regression.

Additional use of conference matrices in screening experiments can be found in [15].

The link for the matrices in **Tables 2** and **3** is at https://designcomputing.net/Cmatrices/.

Author details

Nam-Ky Nguyen[1*], Stella Stylianou[2], Tung-Dinh Pham[3] and Mai Phuong Vuong[4]

1 Vietnam Institute for Advanced Study in Mathematics, Hanoi, Vietnam

2 School of Science, Mathematical Sciences, RMIT University, Melbourne, Australia

3 VNU University of Science, Vietnam National University, Hanoi, Vietnam

4 Hanoi University of Science and Technology, Hanoi, Vietnam

*Address all correspondence to: nknam@viasm.edu.vn

References

[1] Poorna V, Kulkarni PR. A study of inulinase production in aspergillus niger using fractional factorial design. Bioresource Technology. 1995;**54**: 315-320

[2] Mee RW. A Comprehensive Guide to Factorial Two-Level Experimentation. New York: Springer; 2009

[3] Yao CL, Liu CH, Chu IM, Hsieh TB, Hwang SM. Factorial designs combined with the steepest ascent method to optimize serum-free media for ex vivo expansion of human hematopoietic progenitor cells. Enzyme and Microbial Technology. 2003;**33**:343-352

[4] Plackett RL, Burman JP. The design of optimum multifactorial experiments. Biometrica. 1946;**33**:305-325

[5] Ionin YJ, Kharaghani H. Balanced generalized weighing matrices and conference matrices. In: Colbourn JH, editor. Handbook of Combinatorial Designs. 2nd ed. Boca Raton, FL: CRC Press; 2007. pp. 306-313

[6] Nguyen NK, Stylianou S. Constructing definitive screening designs using cyclic generators. Journal of Statistics Theory & Practice. 2013;**7**: 713-724

[7] Jones B, Nachtsheim CJ. A class of three levels designs for definitive screening in the presence of second-order effects. Journal of Quality Technology. 2011;**43**:1-15

[8] Stylianou S. Three-level screening designs applicable to models with second order terms. In: Paper Presented at the International Conference on Design of Experiments (ICODOE-2011). Department of Mathematical Sciences. USA: University of Memphis Memphis; 2011

[9] Xiao L, Lin DKJ, Bai F. Constructing definitive screening designs using conference matrices. Journal of Quality Technology. 2012;**44**:2-8

[10] Fidaleo M, Lavecchia R, Petrucci E, Zuorro A. Application of a novel definitive screening design to decolorization of an azo dye on boron-doped diamond electrodes. International Journal of Environmental Science and Technology. 2016;**13**:835-842

[11] Nguyen NK, Kenett RS, Pham TD, Vuong MP. Recent development on D-efficient mixed-level foldover designs for screening experiments. In: Pham H, editor. Springer Handbook of Engineering Statistics. 2nd ed. London (accepted): Springer; 2022

[12] Jones B, Nachtsheim CJ. Definitive screening designs with added two-level categorical factors. Journal of Quality Technology. 2013;**45**:121-129

[13] Nguyen NK, Pham TD, Vuong MP. Constructing D-efficient mixed-level foldover designs using Hadamard matrices. Technometrics. 2020;**62**:48-56

[14] Nguyen NK, Pham TD, Vuong MP. Multiway blocking of designs of experiments. Statistics and Applications. 2021;**19**:1-9

[15] Stylianou S. Foldover conference designs for screening experiments. Communication in Statistics—Theory and Methods. 2010;**39**:1776-1784

Chapter 7

QSRR Approach: Application to Retention Mechanism in Liquid Chromatography

Jovana Krmar, Bojana Svrkota, Nevena Đajić,
Jevrem Stojanović, Ana Protić and Biljana Otašević

Abstract

One-factor-at-a-time experimentation was used for a long time as gold-standard optimization for liquid chromatographic (LC) method development. This approach has two downsides as it requires a needlessly great number of experimental runs and it is unable to identify possible factor interactions. At the end of the last century, however, this problem could be solved with the introduction of new chemometric strategies. This chapter aims at presenting quantitative structure–retention relationship (QSRR) models with structuring possibilities, from the point of feature selection through various machine learning algorithms that can be used in model building, for internal and external validation of the proposed models. The presented strategies of QSRR model can be a good starting point for analysts to use and adopt them as a good practice for their applications. QSRR models can be used in predicting the retention behavior of compounds, to point out the molecular features governing the retention, and consequently to gain insight into the retention mechanisms. In terms of these applications, special attention was drawn to modified chromatographic systems, characterized by mobile or stationary phase modifications. Although chromatographic methods are applied in a wide variety of fields, the greatest attention has been devoted to the analysis of pharmaceuticals.

Keywords: liquid chromatography, machine learning algorithms, molecular descriptors, QSRR model building and validation, analyte's retention predictions

1. Introduction

One of the most widely applied analytical techniques in a broad variety of application areas is high-performance liquid chromatography (HPLC). It stands out due to its high precision, efficacy, and robustness. Despite its undeniably good aspects, the susceptibility of analyte's retention to a diversity of experimental setup parameters makes HPLC method development a time-consuming and expensive process. Unfortunately, the selection of an appropriate combination of chromatographic conditions related to both a stationary and a mobile phase, as a starting point for the analysis of a particular drug chemical entity, is often done using a trial-and-error approach [1].

IntechOpen

At the same time, one of the major goals of contemporary chromatographic analysis is to efficiently identify optimal working conditions for a better success rate in the method development. Luckily, a tailored, pragmatic approach denoted as quantitative structure–retention relationship (QSRR) modeling was introduced [1, 2]. With the assistance of computerized statistical methods, QSRR models are supposed to mathematically relate the molecular structural properties with the chromatographic response of a drug generated within a set of defined experimental conditions. The molecular structure encodes its physicochemical information in the form of numerical quantities denoted as molecular descriptors. This approach offers great assistance in understanding the analyte's chromatographic behavior and enables the discovery of physicochemical processes involved. As expected, statistical QSRR studies are, therefore, recognized as a supreme chemometric approach leading to the timely enhanced, high-quality separation, and efficient analytical method development [1, 2].

QSRR models are commonly associated with the retention prediction of a new and non-analyzed compound. However, QSRR models are much more useful since they are applied in revealing the molecular descriptors with the greatest retention predictive potential as well as in revealing the mechanisms that govern the separation in a specific chromatographic system on a molecular level. Based on a reliable QSRR model that accounted for different sets of chromatographic data within the same type of stationary phase (e.g. reversed-phase (RP)), the quantitative comparison of chromatographic columns can be achieved [1, 3]. The additional value of the same data refers to the direction where to look for a chromatographic column with equivalent performance and orthogonal selectivity as well as to upgrade chromatographic performances that are the most responsible for retention parameters inclusive of a short overall run time [4]. Besides all the aforementioned, many authors assert that the retention in an HPLC system, especially in RP- and micellar chromatographic systems, can be closely related to the biological activity of a drug. This can be understandable in terms of a compound's lipophilicity and pKa value because its chromatographic distribution between stationary and mobile phases is highly similar to its bodily distribution between the cell membrane and intracellular or extracellular fluids. As a result, the chromatographic data can be related to the description of biological processes of drug absorption, distribution, and excretion as well as drug-receptor interactions. Looking at the QSRR study within these wider frames, this approach can be used as a valuable in silico method for the prediction of the analyte's lipophilicity and biological activity of potentially new drug molecules. In such a manner, the utilization of less effective experimental methods and animal models can be reduced [3].

Because of their wide applicability, the QSRRs methodologies have been quite extensively studied over the past two decades. The first article, in which Tamf and Kamlet mention QSRR in a similar context known nowadays, dates from 1977. However, an intense interest in this topic has arisen over the last two decades after the work of Roman Kaliszan [5]. The first theory used to describe chromatographic retention was the theory of linear free-energy relationships (LFER), according to which the analyte's retention parameters reflect the free-energy changes associated with the chromatographic distribution [6]. In that regard, a chromatographic column is recognized as a "free-energy transducer," which translates the chemical structure differences of compounds into quantitative differences in the retention parameters. In order to provide the proper knowledge about a chromatographic system, a relatively large set of reliable input data, coming from a group of structurally heterogeneous compounds and retention data, is needed. The early introduced QSRR models were based on a priori selection of a small set of structural descriptors derived from a molecular formula or a molecular graph reflecting physicochemical properties. These sets of structural descriptors are well known to chemists since they originate from the accepted theories of chromatographic separation and the interpretation of fundamental intermolecular interactions [7]. Since the representation of the separation process solely in terms of intermolecular interactions is questionable, an approach based on linear solvation energy relationships (LSER) was introduced by Abraham [6, 7]. He pointed out a new notably expanded set of molecular descriptors indicating the difference in interactions as a consequence of the solvent properties of the mobile and stationary phase. In parallel with these considerations, to provide reproducible quantitative input chromatographic data, two main methodological directions may be observed in the literature. The retention data can be determined under the same experimental condition or by varying chromatographic conditions, such as mobile phase compositions, flow rates, column temperatures, etc. [1]. The latter approach, also known as mixed QSRR modeling, is found to be advantageous. It enables the recognition of patterns in analyte's retention changes within observed experimental ranges of chromatographic parameters and consequently an in-depth understanding of complex chromatographic systems. In addition to proper input data, statistically significant and physically meaningful QSRR modeling finally relies on solid mathematical analysis. The usual technique for the mathematical description of correlations between all gathered data is multiple linear regression (MLR). However, the advances in liquid chromatography (LC) and an increase in the amount of chromatographic data generated over time make the conduction of a QSRR study difficult to handle traditionally. In that regard, QSRR models have shifted from a priori selection of simple descriptors and traditional regression analysis to the generation of a large pool of molecular descriptors and machine learning algorithms (MLAs) based on linear and/or nonlinear regression analysis [1]. For the sake of obtaining chemically valid interpretations, useful and reliable QSRR models demand a selection of the most informative and predictive descriptors among often mutually correlated ones. Therefore, the need for suitable selection techniques for input information data emerged accompanied by QSRR model validation strategies used to evaluate model prediction performance [2]. High-performance calculations at all the stages have made the process of LC method development more efficient and sustainable. They have also improved the fundamental knowledge of the separation processes. In accordance with numerous benefits, the anatomy of the QSRR modeling is reviewed below in conjunction with the guidance of modern requirements and tendencies.

2. QSRR workflow: a detailed walkthrough

2.1 Molecular descriptors

The power of QSRR comes from the characterization of compounds via molecular descriptors (MDs) that depict the physicochemical information of molecules in a numerical manner. The concept of MDs has come a long way in the last 50 years as it witnessed constant progress in computational chemistry. The accompanying advances in hardware enabled the calculation of over 5000 descriptors for a single molecule [8, 9]. Depending on the classification criteria, molecular descriptors can be divided into several groups. Some descriptors are obtained experimentally, while the others are purely theoretical. According to the data type, molecular descriptors can be Boolean, integer, real, vector, etc. According to the structural dimensions (D), on the other hand, molecular descriptors can range from 1D to 6D [10, 11]. Based on the references outlining their application in QSRR studies, the extensively used descriptor-calculation software are AlvaDesc [12–14], Dragon [15–19], Molinspiration Cheminformatics [20, 21]. and Chem3D Ultra [19]. The latest and freely available software, PaDEL-Descriptor [14, 18] and Mordred [22, 23], allow MDs to be computed under open science practice representing a valuable addition to the palette of commercial software. For more simple descriptors (e.g. compositional or topological descriptors), a simplified molecular input line entry system string (SMILES) or a 2-D map can be used to represent molecules under study. If descriptors give more information, molecular geometry has to be determined prior calculation process. The accuracy of the most descriptors subsequently depends on the method used to build a 3-D molecular structure. Given a variety of the computational methods used for optimizing the geometry of analytes for QSRR studies and the availability of resources, researchers can opt to perform empirical force field methods (e.g., molecular mechanics), semi-empirical optimization (e.g., AM1, PM3), or sophisticated *ab initio* calculations (e.g., Hartree-Fock and Density Functional Theory) [24, 25]. In an interesting study, Amos et al. investigated how different levels of theory for structure optimization contributed to the QSRR outcome. The sum of ranking differences (SRD) showed that a fast and rational method of structure optimization shared the results with time-consuming and expensive calculations in terms of the final accuracy of the QSRR model. Moreover, the solvent correction did not reduce the mean absolute error of QSRR predictions. The authors carefully explained these unexpected findings in the context of an error inherent in the Dragon descriptor calculation process [26].

2.2 Feature selection

A small set of predictors (i.e., input variables or factors) with well-known physicochemical meaning can be pre-defined when modeling separation in systems with fully elucidated retention mechanisms. For complex chromatographic modes, such as micellar liquid chromatography (MLC) and mixed-mode liquid chromatography (MMLC), a priori attribute selection may compromise the accuracy of QSRR predictions making it a poorly acceptable strategy for retention modeling [27, 28]. Alternatively, a large set of independent variables can be formed; the most significant attributes can be extracted from it and used to build a model for retention time prediction. Clearly, the predictive ability of these models depends on the efficiency of the mathematical algorithm used to select predictor variables [7, 29].

The choice of the most informative features for a particular regression problem poses one of the main challenges in machine learning (ML). Determining the appropriate method of variables selection, in this regard, has been an interesting topic in a broad range of domain applications, including studies for which the datasets with hundreds or thousands of attributes become available along with the development of molecular modeling software. Faced with plenty of noisy and irrelevant features, contemporary QSRR studies call for variable selection without exception. The purpose of variable selection methods is to handle space dimensionality by discarding the features that are redundant and irrelevant in predicting endpoint values. A feature is irrelevant if it is unpredictive for the dependent variable or response. A reduction is needed if it is highly correlated with other features. The adoption of feature selection techniques ultimately avoids overfitting, improves a model's predictive power, and enhances an understanding of the underlying patterns preserved in data. A decreased computational burden placed on modeling techniques as well as easier data visualization happens to be additional benefits associated with feature selection techniques [29–31].

In a typical MLA-empowered QSRR pipeline, a minimal feature subset is determined after the pre-processing of raw data and before the modeling. Among various techniques, MLR, genetic algorithm (GA), and Relief method have been quite eagerly used in QSRR studies [31, 32]. Other important feature selection methods are least absolute shrinkage and selection operator (Lasso), artificial neural network (ANN), and random forest (RF) [24, 27, 33]. The last two algorithms will be discussed later in the text, as part of Section 2.4.

2.2.1 Multiple linear regression

MLR founds a linear relationship between a dependent variable and two or more independent variables (regular attributes). It is basically the extension of the Ordinary Least Squares (OLS) method.

The general MLR model can be written using Eq. (1):

$$y_j = \beta_1 x_{j1} + \beta_2 x_{j2} + \ldots + \beta_n x_{jn} + \beta_0 + \varepsilon \tag{1}$$

where y_j is a dependent variable, x_j are independent variables, β_n are slope coefficients for each predictor, β_0 is an intercept, and ε refers to a model's error term.

In the OLS method, the slope coefficients that minimize the loss function come from Eq. (2):

$$\sum_{j=1}^{k}\left(y_j - \hat{y}_j\right) = \sum_{j=1}^{k}\left(y_j - \left(\beta x_j + \beta_0\right)\right)^2 \tag{2}$$

The use of MLR makes sense only if: a) there is a linear relationship between predictors and dependent variable, b) the correlation between variables is not too high, c) the instances are chosen randomly from the population, and d) the residuals are normally distributed. MLR estimator is burdened with a great variance, especially in the cases, where the number of attributes approaches the number of observations [27].

2.2.2 Least absolute shrinkage and selection operator

Lasso regression is one of the most popular regularization methods for selecting significant independent variables. The concept of regularization has been introduced to avoid overfitting in MLR modeling. In brief, the regularization refers to adding a "penalty" term to the best model built upon a training dataset and to achieve a smaller variance and control the influence of the predictor variables over the response. In Lasso regression, this is done by penalizing the absolute value of the magnitude of coefficients (Eq. (3)).

$$LASSO = \sum_{J=1}^{K}\left(y_j - \sum\nolimits_n x_{jn}\beta_n\right)^2 + \lambda\sum_{k=1}^{p}|\beta_k| \tag{3}$$

In Eq. (3), λ is the tuning parameter that controls the amount of shrinkage. If λ is large, the slope coefficients are penalized highly toward 0, and more features are eliminated. If λ is 0, all features are considered and the residual sum of squares criterion is applied. As λ increases, the bias increases. Otherwise, the variance increases [27, 34].

2.2.3 Genetic algorithm

GAs are methods that generate a solution for optimization and search problems by simulating the mechanism of natural selection and the survival of the fittest. In the initial stage, the GA creates a random population of chromosomes. Each chromosome, usually represented by a binary string, encodes a potential solution to the problem under study. In the case of feature selection, individual chromosomes make up a random subset of variables, where the presence or absence of a variable in the chromosome is denoted by 1 or 0, respectively. Using individuals in the current generation, the GA creates a sequence of new populations. To achieve this goal, the algorithm first evaluates each chromosome of the current population by determining its fitness value. The fittest individuals are selected to pass their genes to the next generation. Offsprings are, in fact, produced by subjecting the selected parents to crossover (gene exchange) and mutation (gene change in individuals). In addition, some of the population's members with the best fitness values are chosen as elite children and added directly to the next population. The subsequent generation is formed after children with inherited good characters replace the current population of parents. The GA loops until one of the stopping criteria is met (e.g. a predefined number of generations). The flowchart (**Figure** 1) outlines the main GA steps.

In terms of the prediction accuracy of constructed QSRR models, the GA showed superiority in selecting the most relevant features compared to other variable selection methods [18, 35]. Lately, the GA has been used for the non-polynomial hard problem of feature selection [36], the selection of molecular descriptors for localized QSRR models [37], and the development of a QSRR model intended to improve the structural annotation of triterpene metabolites in an LC-HRMS system [38].

2.2.4 ReliefF

In this method, each attribute is assigned a relevance weighting according to its ability to distinguish between class labels. Attributes with weight above the user-defined threshold τ are considered significant and included in the set of selected features. The underlying principle is that the instances belonging to the same class should be closer than those of different classes. The algorithm cycles over j training cases (R_i) that are chosen by the user. First, n dimensional weight vector $W$of zeros is initialized. Then, the target instance R_i is selected at random and the distances between it and its two nearest neighbors, namely, *nearestHit* (the closest instance with the same class) and *nearestMiss* (the closest instance with the opposite class) are calculated. Feature weight W is updated so that more weight is assigned to attributes that distinguish an instance from neighbors of different classes (Eq. (4)). After j cycles, each element of the weight vector is divided by j, giving rise to the relevance vector [27, 30, 31].

$$W_i = W_i - (R_i - nearestHit)^2 + (R_i - nearestMiss)^2 \quad (4)$$

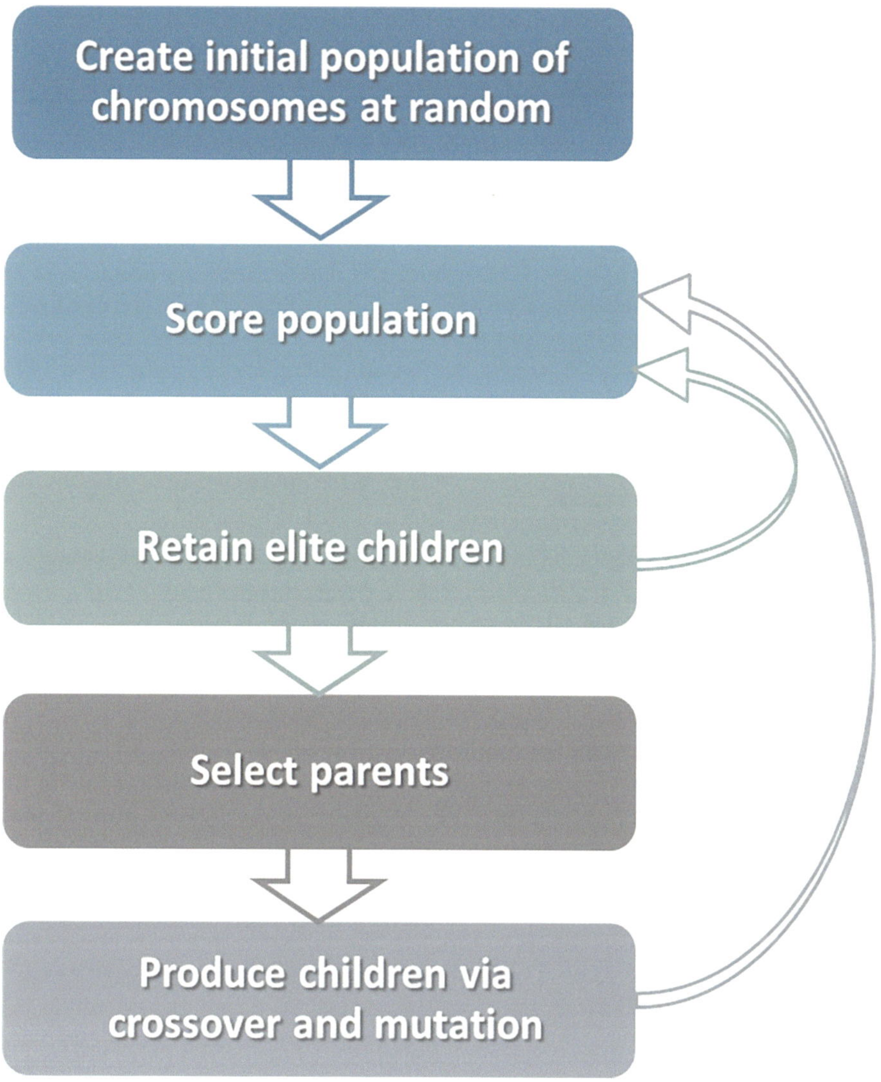

Figure 1.
GA flowchart.

Originally, the Relief algorithm has been intended for classification problems and could be fooled by an insufficient number of cycles. Nowadays, it has been adapted for predicting continuous decision variables and as such is being used for QSRR studies (e.g. to predict retention parameters). The differences between Relief, ReliefF, and RReliefF are presented in detail in [39].

2.3 Response transformation

When implementing supervised algorithms, it is a good practice to examine data distribution. The distortion of the symmetry of normal distribution around its mean is denoted as skewness. A general impression of skewness can be gained by drawing a histogram or computing the skewness coefficient. If the distribution's shape has one peak and a long tail on the right side of the curve, the distribution is positively skewed.

In contrast, the distribution has a negative skew if a long tail is on the left side of the curve. In numerical terms, the skewness for a normal distribution is (approximately) 0. Negative coefficients are related to negative skewness and vice versa. The coefficient values between – 0.5 and + 0.5 indicate moderately skewed data, and if they are less than −1 or greater than + 1, the distribution is highly skewed. A highly skewed dataset can contaminate a model's predictive performance because the algorithm has to deal with scattered endpoints at extreme values. In the case of right-skewed data, for instance, MLAs are likely to predict points with lower values better than those with high values. Therefore, skewed distribution is one of the major obstacles to the application of MLAs to real-world data and should be addressed prior to the modeling. A common strategy for dealing with skewed variables is to transform them. Logarithmic, square root, and cube root transformations are recommended when data follow the power-law distribution, while in the opposite case, it is better to opt for square, higher powers, or cube root transformations [27, 40, 41].

2.4 Model building techniques

The choice of regression technique for correlating molecular descriptors and chromatographic conditions with a chromatographic parameter has a huge impact on the performance of any QSRR model. Due to its simple and explainable character, MLR received considerable attention in mechanistic research long ago [24]. However, if researchers amass vast troves of data and cannot make sense of it in a reasonable amount of time, the process is the main candidate for modeling through more sophisticated MLAs. MLAs fall under the umbrella of artificial intelligence and can process and understand data faster. These algorithms learn to resolve issues by drawing firm conclusions from observation data they are supplied with. Along with improvements in technology and computing power, QSRR can take advantage of machine learning in a fundamental and practical manner. By acknowledging nonlinearity in LC data, MLAs play an important role in the accuracy of property predictions. However, no currently available MLA can deliver optimal performance for every modeling task. A variety of MLAs should be used before selecting a particular regression technique. The common MLAs are ANNs, support vector regression (SVR), and ensemble methods [42].

2.4.1 Artificial neural networks

ANN is a series of machine learning algorithms that mimic the process of natural thinking by making experience-based decisions. Modeled on the human brain, the ANN refers to a massive composition consisting of some primitive processing elements (i.e. artificial neurons). Most operative neural nets are constructed by grouping neurons into layers. An individual neuron might be connected to several nodes in the layer beneath, i.e., above it. Data passing through layers in only one direction makes up a feedforward neural net (or multi-layer perceptron). Apart from the layers, the main components of ANN include the adaptive coefficients –weights, assigned to each of the connections between the layers, as well as the transfer functions, which convert received raw data into output. The transfer functions, learning rules, and architecture itself define the behavior of each ANN [43]. When a neural net is being trained, all of the weights are first randomly assigned to synapses between neurons. Then the input-output pairs of data are fed to the net in an attempt to train an algorithm to recognize the underlying patterns between variables. This strategy pertains to the process of supervised learning. In a supervised feedforward backpropagation algorithm, the training is performed by comparing the processed signals with the desired outputs and adjusting the inputs' weights until the margin of error is minimal. Herein, the weights are updated in the steepest descent fashion. Higher weights are attributed to the inputs that contribute the most to achieving the right target [44, 45].

Neural nets are a valuable tool for analytical R and D due to the ability to learn nonlinear relationships encountered between predictors and dependent variables in most corresponding systems. Contemporary applications of ANN in the pharmaceutical sciences are broad, ranging from interpretation of analytical data to drug design. Over the past decade, there has been an impressive increase in the number of publications on QSRR studies that used ANN as a modeling technique. In particular, the single-hidden layer neural nets provided a satisfactory level of prediction accuracy [46–51]. After the improvement in computer power and the rise of big data, ANNs began to flourish in the form of deep learning (DL) algorithms [52–55]. Deep neural networks are the ones that have more than one hidden layer. With each additional layer, the DL algorithm can model increasingly complex relationships. As compared to other ML techniques, ANN architecture is characterized by great flexibility and can process raw data and automatically extract a set of the most informative features. Unfortunately, the DL is not free of limitations; in general, these algorithms are data-hungry and require massive training sets. The question to be raised, in that respect, is whether the analytical domain can provide big data without losing valuable resources [56, 57]?

2.4.2 Support vector regression

SVR is another promising machine learning algorithm that acknowledges the nonlinearity in data. It is built on the principles of statistical learning and the concept of constructing a line (or hyperplane in high-dimensional space) that fit the data. Among an infinite number of possible solutions, SVR finds a hyperplane with the greatest distance to the nearest training instances. Finding such a hyperplane is based on minimizing the l2-norm of the coefficient vector, w (Eq. (5)), while the absolute error between the target y_i and predicted values are set to be less than or equal to a specified margin, ε (Eq. (6)).

$$\min \frac{1}{2}\|w\|^2 \tag{5}$$

$$|y_i - w_i x_i| \leq \varepsilon \tag{6}$$

In Eq. (6), x_i is the i-th input point in the input space (a feature) and w_i is its associated coefficient. The maximum error ε is tuned to gain the predictive ability of the built model satisfactorily.

For the endpoints that reside outside the ε-tube, deviation from the margin is represented by the slack variable, ξ. Term C is added to penalize these points in comparison with those either above or below the hyperplane. With respect to these deviations, the objective function and its constraint are given in Eqs. (7) and (8), respectively.

$$\min \frac{1}{2}\|w\|^2 + C\sum_{i=1}^{n} \xi_i \tag{7}$$

$$|y_i - w_i x_i| \leq \varepsilon + |\xi_i| \tag{8}$$

The SVR hyperplane is constructed after the inputs are mapped into a space of higher dimension(s) than the original using the kernel function (e.g., polynomial, splines, radial basis function, etc.). Then, using a simple linear function, the SVR helps predict the target value. By projecting the optimal hyperplane back into the input space, it takes on a nonlinear form. Due to its remarkable generalization ability, the SVR has gained popularity in QSRR studies [31, 44, 55, 58–63]. In most publications, the empirical performance of SVR matches with or is considerably better than the performance of other MLAs studied.

2.4.3 Ensemble learning algorithms

In ensemble learning, algorithms with high bias or too much variance (so-called weak learners) are merged to produce the most popular result. The underlying idea of aggregating predictions is to create a much more accurate and robust model. Bagging (also known as bootstrap aggregation) and boosting are the most prominent classes of ensemble methods [64].

Weak learners that are used widely in ensemble learning are decision trees (DTs). DTs are nonlinear machine learning techniques that can handle either regression or classification tasks. They are simple, intuitive, and can deal with missing values and large datasets with elegance. The classification and regression tree (CART), introduced in 1984, is a typical DT algorithm [65]. It is presented as a tree-shaped diagram containing a set of nodes and branches growing downwards. This topology gives the idea of a binary and hierarchical algorithm that adopts the recursive partitioning method. It is an iterative procedure that seeks to find the best split (the best splitting feature and the best input data) at each step. Performance metrics, e.g., Gini index, information gain, or error rate, are utilized to assess the quality of the split. Fundamentally greedy nature and poor ability to cope with the penalties on tree complexity (while growing the tree) are the main disadvantages of the top-down approach. Pruning is done to prevent an overfitting phenomenon [66].

2.4.4 Random forest

RF was introduced as a DT-based ensemble in 1984 [65]. It is a collection of unpruned DTs (grown to the maximum extent) that are trained by the bagging method. In bagging, base models are grown on bootstrapped subsets of the data and the individual predictions of all base models are averaged to get the final output. As a result, the ensemble model has less variance than its building elements. While sharing the main idea with bagging, the RF adopts an additional level of randomness – each node of each tree deliberately takes into account only a random subset of features (e.g., the square root of a number of descriptors) for the splitting procedure. An RF model benefits from this tactic in terms of efficacy. In addition, it is important to mention that the internal validation is built into the forest growth. According to the concept of bootstrapping, some of the data are omitted from the samples intended for tree growth, while the others are repeated in the samples. The former is denoted as Out-Of-Bag (OOB) data. Given the fact that the OOB sample is not included in the tree fitting, it is used to estimate the model error. Usually, it makes up to one-third of the available data, while the other two-thirds of the data is used for training. In order to achieve a small OOB error, it is necessary to optimize the number of base models and the size of a subset of features [66, 67].

In QSRR studies, the RF algorithm is readily used as a modeling technique [42, 55, 66, 68] as well as a feature selection method [69, 70]. The latter is due to the ability of the algorithm to quantify the importance of variables under study. The importance of each feature is determined by observing a change in prediction error when the OOB set for that feature is permuted (and the other features are kept constant).

2.4.5 Gradient boosted trees

Gradient boosted trees (GBT) is an extremely powerful ensemble algorithm based on boosting and gradient descent approach. Unlike the bagging, which combines weak learners in parallel, the boosting merges base models linearly. The focus here is especially on shallow DTs that have low variance and high bias.

A correlation between base models (arising from the same data) is precluded by an incremental change of the training set. This is done by assigning weights to each example. Initially, all weights are set to be equal and the first decision tree is trained on the original dataset. Accurately predicted instances have their weights decreased, while the others have their weights increased. The trees that enter the ensemble in subsequent iterations are thus applied to the reweighted data and their goal is to correct the errors made by the previous model. Boosting, which decreases the bias of individual base models, is viewed as one of the groundbreaking concepts introduced in ML over the last decades. The GBT algorithm minimizes a loss function via a gradient descent procedure. The predictive power of the GBT ensemble correlates with the number of base models and the size of learning rate. A larger ensemble will very quickly over-fit, while a combination of too few DTs might lead to poor predictive performance. Lower values of learning rate (a parameter that controls the length of incremental step) may resolve the problem of overfitting, but a prolonged convergence toward the solution can place a lot of computational burden on the model in question [71, 72]. Due to the ability to create highly accurate QSRR models (and the fact that it quite often outperforms many other regression algorithms), the GBT is popular in analytical R&D. The successor to the gradient boosting, regularized gradient boosting (i.e., XGBoost), is increasingly used to provide state-of-the-art solutions to many LC challenges as it yields improved generalization capabilities and better avoids over-fitting [27, 32, 42, 73, 74].

2.5 QSRR model validation

It seems that it is feasible to build mathematical models that fit the data very well. But, there is still a possibility that it may happen due to chance correlations or overfitting. In that case, the models are not considered appropriate for their intended application. Therefore, proper statistical validation of models is of great importance in QSRR studies. The two main concepts, denoted as internal and external validation, will be discussed herein. The internal validation procedures include leave-one-out (LOO) and leave-many-out (LMO) cross-validation (CV), y-randomization, and bootstrapping.

2.5.1 Leave-one-out cross-validation

LOO-CV is performed by excluding each sample (compound) once and building a model with the remaining data and predicting the value of the response for the eliminated sample. Due to the presence of repetitive cutting data set activities, the

LOO is also known as rotation estimation and jack-knife validation method. This approach indicates that the eliminated sample serves as a temporary test set taken from the overall training set. Each cycle of this repetitive procedure is followed by calculating the differences between experimentally observed response values and estimated (predicted) ones by the model. These values are afterward included in Eqs. (9) and (10) corresponding to the root mean square error of CV (*RMSECV*) and the cross-validated correlation coefficient (Q^2), respectively. Finally, the model predictive performances are inspected by the values of the root mean square error of calibration (*RMSE*, Eq. (11)) and the overall CV correlation coefficient value, calculated for the whole original dataset as the average value of Q^2 from each CV cycle [27]. The value of overall Q^2 is usually greater than that of individual Q^2, though a large difference between them (overall Q^2 greater by 25%) indicates that the model suffers from overfitting [75].

$$RMSECV = \sqrt{\frac{\sum \left(y_{experimental} - y_{predicted}\right)^2}{n-1}} \tag{9}$$

$$Q^2 = 1 - \frac{\sum \left(y_{experimental} - y_{predicted}\right)^2}{\sum \left(y_{experimental} - \left\langle y_{experimental}\right\rangle\right)^2} \tag{10}$$

$$RMSE = \sqrt{\frac{\sum \left(y_{experimental} - y_{predicted}\right)^2}{n}} \tag{11}$$

In the aforementioned Eqs. (9)–(11), n stands for the number of samples in the dataset, $y_{experimental}$ are the values of experimentally observed responses and $y_{predicted}$ are the responses calculated (theoretically predicted) based on the built model calculated either from the data used from model development (in case of Q^2 and *RMSECV*) or the original dataset (in case of *RMSE*). The brackets $^{<>}$ are used to point out the use of the average values of experimentally obtained responses.

2.5.2 Leave-many-out cross-validation

To perform the LMO-CV, the initial dataset is divided into blocks of samples; afterward, each block is eliminated once from the model building in each cycle in a similar manner as applied in the LOO-CV. The prediction of response is made for the block under consideration. It should be noted that the blocks may consist of the same number of constituents, but that is not an obligation. The LMO may also have different validation cycles. In that respect, as an example, the original dataset can be divided into 10 parts indicating that each data block accounts for 10% of the data, 10 validation cycles are needed and the respective method may be referenced as 10-fold CV as well. In comparison with the LOO, this procedure is more time-effective. The validation metrics are similar to those presented for the LOO-CV with adjustments in relation to a sample or a block of samples. It is worth mentioning that the same appropriate adjustments must be implemented in the used equations. Also, since there is no truly new compound under consideration within none of the variations of the internal validation procedures, it is advisable to perform as many as possible internal validation tests for the final justification that the model is of good quality, relevant, and suitable for its intended use. This recommendation especially stands in the case of the modeling based on small datasets where any omission of data from the original dataset may lead to the inability to perform the modeling procedure at all [75, 76].

2.5.3 y-randomization

Y-randomization is used to ensure the robustness of the developed model. For example, it can check whether there is a molecular descriptor statistically well correlated with the response value y; but, in reality, there is no cause-effect relationship originating from the physical and/or chemical meaning of a molecular descriptor and the respective retention measurement. The model validation is performed by keeping the so-called X matrix with original unchanged descriptors while the vector of the response values y is randomized or scrambled. Since the new models are built based on the same input dataset but associated with changed (false) responses, it is expected that they are of poor quality as reflected by the values of Q^2 and overall Q^2. Kiralj and Ferreira proposed a detailed overview of the possible Q^2 and overall Q^2 values and their interrelationships according to which the chance correlation may be inspected [77].

2.5.4 Bootstrapping

Bootstrapping procedure suggests the random splitting of a complete dataset into training and test sets several times and the building of respective models afterward. While in the LOO and LMO procedures each sample is excluded from the modeling only once, in the bootstrapping there is an equal chance for a sample to be eliminated once, several times, or even never. The corresponding Q^2 and overall Q^2 validation metrics are calculated and expected to be of high values as well as to oscillate around the real values or values obtained from the LOO-CV of a real model. It should be noted that this validation procedure is affected by a number of splits or resampling as well as the structure or similarity between the training and test sets [77].

2.5.5 External validation

The predictive power of a QSRR model is evaluated by the external validation, with *model blind* samples (compounds), meaning that these samples were not previously *seen* by the model or used for model development. Therefore, the extraction of an external validation test set from the original data is required, and a proper selection of the size and type of these data is of crucial importance for a successful validation process. Usually, this subset covers 15–25% of the original dataset [78, 79]. Although the external validation test set is a golden standard of the QSRR models' prediction properties, there is a concern about the relatively small size of the external test set in comparison with the LOO- or LMO-CV where the whole dataset acts as a test set in some moment. At the same time, the consideration of the similarity between the training and external validation subsets is of utmost importance by means of similar variable ranges and distribution. The common trend indicates that a greater similarity between these subsets leads to a decrease in prediction errors [4]. Finally, more than one splitting of the dataset into modeling and external validation test sets is also advisable [80].

The statistical parameters for model evaluation include the multiple coefficients of determination of external validation, also called *predictive* R^2, and the root mean square error of prediction (*RMSEP*). The values of these parameters can be calculated using Eqs. (10) and (11), taking into consideration that all data correspond to the external test set solely. Another valuable indicator of the model's predictive performance is the Pearson correlation coefficient of prediction (*R*), which is used to reflect a correlation existing between the experimentally observed responses and the responses predicted by the model. It is expected from the value of *R* to be maximally close to 1. The parameter *R* can be calculated by Eq. (12) [77].

$$R = \frac{\sum\left(\left(y_{experimental} - \left\langle y_{experimental}\right\rangle\right)\left(y_{predicted} - \left\langle y_{predicted}\right\rangle\right)\right)}{\sqrt{\sum\left(y_{experimental} - \left\langle y_{experimental}\right\rangle\right)^2}\sqrt{\sum\left(y_{predicted} - \left\langle y_{experimental}\right\rangle\right)^2}} \tag{12}$$

2.5.6 Detection of sources of prediction errors

Apart from the inspection of the statistical parameters computed from the respective validation procedure, to assure the quality and practical usefulness of QSRR models, it is worth getting insight into the possible sources of prediction errors (residuals). In that respect, besides the calculation of *RMSE,* which uses the same units as the response, it is useful to express it in percentages. By analyzing the value of *RMSE (%)*, the magnitude of the prediction error concerning the mean of actual experimentally observed values is clearer for understanding. Another benefit of this is the possibility to detect outliers i.e., the samples for which the predicted values are too distant from the mean of the experimentally observed values. The outliers differ significantly from all other observations due to the exceptional chemical nature or chromatographic behavior of a compound and may occur in the test dataset as well as in the training dataset. Since their values lie outside the overall usually normal distribution of a dataset, it is quite obvious that the outliers can cause serious problems when it comes to the development of reliable and statistically stable QSRR models. Based on the number of outliers and the intensity of their distinction from other data points in a dataset (soft or influential outliers), the model predictive ability and/or model statistical stability may be brought into question [44]. It is recommended that the outliers should be removed from a dataset before proceeding with model development and analyzed for the origin of possible errors [77, 78, 81]. For the sake of building models of suitable quality, various methods for outlier detection immerged among which some are based on visual analysis of scatter plots, histograms, Box plots, and the others on the calculation of Z-score and interquartile range. More sophisticated methods propose so-called acceptable error windows and unambiguous cut-off limits for applicability domain margins while considering the chemical structural diversity of compounds in a dataset, standardized residuals of predictions and a specific leverage (structural) value of each compound (OTrAMS method), a standard deviation of predictive residuals and a mean of predictive residuals (Monte-Carlo sampling method). More detailed information on the use of later outlier detection methods was provided by Aalizadeh et al. [59].

2.5.7 Definition of model applicability domain

In addition to the statistical model assessments, the predictive power of a robust and validated QSRR model must be expressed in terms of the applicability domain. The model interpretability is affected by its characteristics as well. This domain refers

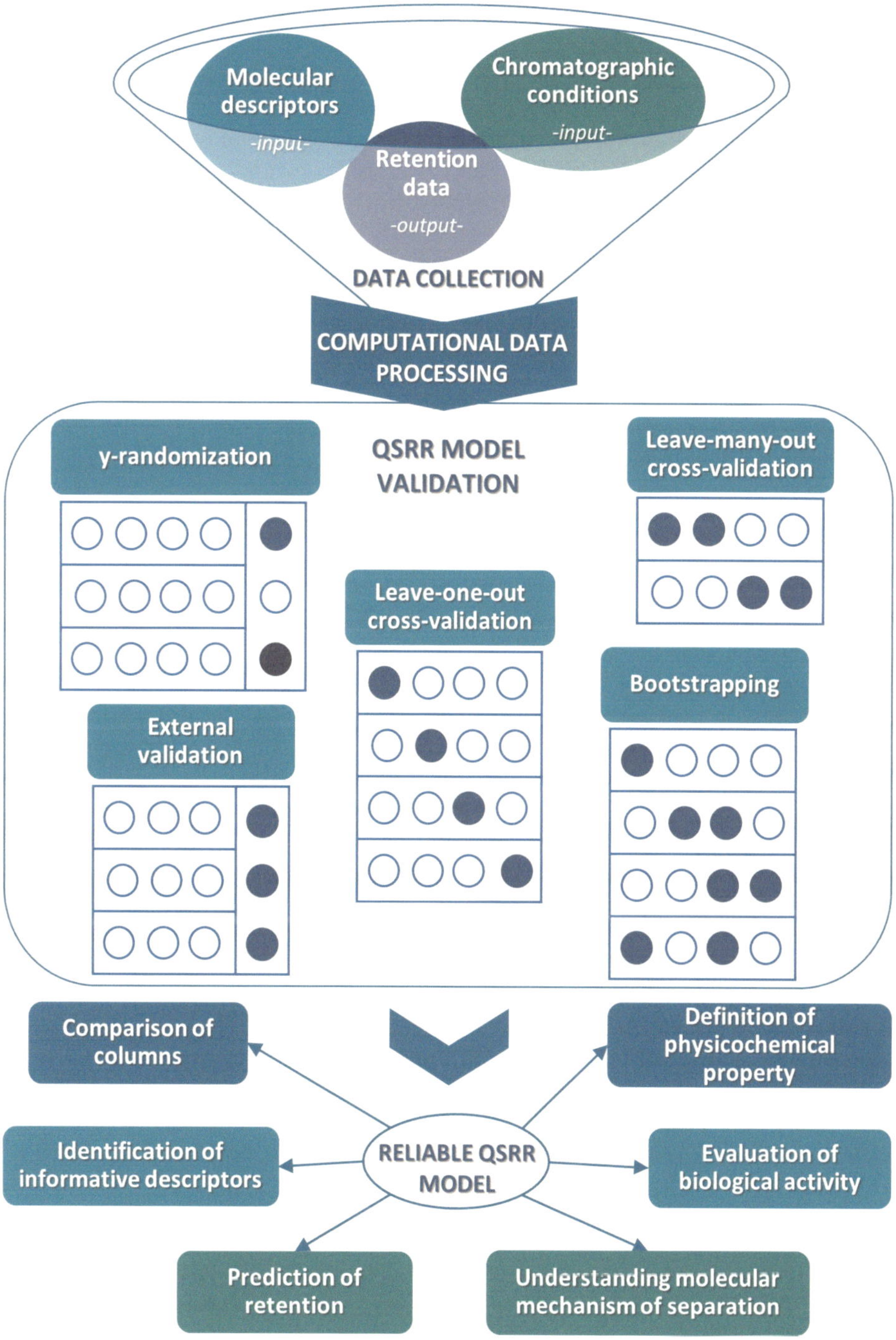

Figure 2.
QSRR workflow.

to a theoretical space defined by a range of the molecular descriptors of compounds used for model training purposes and respective chromatographic conditions as well as a range of the modeled responses. It is obvious that the applicability domain strongly reflects the physicochemical and structural properties and chromatographic behavior of compounds from a training set. In order to make the best response predictions, the training set must be similar to the target molecule [24, 44, 75]. The aim of narrowing the space for making predictions actually serves to avoid unjustified and inaccurate model extrapolations. The dedicated approaches for the definition of applicability domain based on the range of response variables or the range in the descriptor space (geometrical methods and distance-based and probability density distribution-based methods) were thoroughly described by Roy et al. [82, 83]. As the issue is closely related to the applicability domain, the same authors elaborate the strategy for a proper selection of data to be introduced in the training and/or test dataset out of the original dataset as well [83]. It is perfectly reasonable to state that the lack or poorly conducted selection of compounds increases modeling errors and calls into question the success in all predefined QSRR modeling goals or application areas.

After summing all the previous considerations into a graphical presentation, the QSRR flowchart may look like the one in **Figure** 2.

3. Application of statistical QSRR model in complex HPLC techniques

The application of the QSRR approach is directly driven by its definition. As the QSRR represents a mathematical relationship between molecular retention behavior and its properties inherent in molecular structures (molecular descriptors), they are primarily used to predict the retention behavior of molecules omitted during model development. In addition, it can be used to single out important features, by which the retention behavior is governed and it is possible to gain insight into the retention mechanisms. It can also be applied for stationary phase characterization or their comparison in terms of separation characteristics [5]. In some cases, they can provide drug or xenobiotics classification or an assessment of their bioactivity [2]. By incorporating experimental parameter values into a QSRR model, their application can be expanded on HPLC method development and optimization [84].

Since various highly adaptable mathematical tools are suitable for structuring statistical QSRR models, the QSRR approach shows compatibility with a broad spectrum of HPLC properties. Although it has a place in the modeling of conventional unimodal HPLC, which was discussed in more detail by Haddad et al. [84], it is also a valuable tool in the case of defining more complex HPLC systems. Complicated molecular retention patterns are often generated from mobile or stationary phase modification. Taking into consideration such HPLC system modification, the predictive abilities of QSRR can not only reduce experimental requirements but also provide a deeper insight into the retention mechanism. The following section is not a comprehensive literature review but rather a demonstration of the beneficial properties of QSRR used for characterizing complex HPLC systems applied for the analysis of small molecule substances.

3.1 QSRR approach for HPLC with mobile phase modifications

Increasing the retention of poorly retained analytes in RP-HPLC is often achieved by modifying mobile phase properties. The addition of modifiers can provoke changes in the retention behavior by imposing an additional equilibration process.

3.1.1 Ion-interaction chromatography

Compromised retention of basic solutes can be promoted by introducing ion-interaction agents into the mobile phase. Ion interaction chromatography (IIC) involves a series of equilibration processes between chromatographic phases and analytes, which necessitates the understanding of the separation process [85, 86]. An IIC system, with added chaotropic salts, was assessed by Čolović et al. [63]. A mixed QSRR-SVR model was developed based on the retention data of 34 analytes as independent variables were selected i.e., three mobile phase parameters (concentration of NaPF6, pH, and acetonitrile content) and four molecular descriptors (Branching index EtaB with ring correction relative to molecular size (ETA_EtaP_B_RC), calculated octanol/water partition coefficient (XlogP), 3D topological distance-based autocorrelation – lag 9/weighted by polarizabilities (TDB9p) and radial distribution function – 045/weighted by relative polarizabilities (RDF45p) descriptor). The importance of analytes' steric effects and voluminosity were indicated by ETA_EtaP_B_RC, while XlogP implied the significance of hydrophobicity, which was in line with the RP retention mechanism. However, TDB9p and RDF45p indicated the participation of electrostatic interactions during the retention process. Thus, the hypothesis on the complementarity of the analytes' electronic structure and the electrical bilayer created in the stationary phase was supported.

3.1.2 Micellar liquid chromatography

In MLC, a modification of mobile phase features is attained by adding surfactants. When surfactants are present at a concentration above the critical micellar concentration, micelle formation occurs. Surfactant molecules can coat the stationary phase as the absorbed monolayer. Moreover, surfactant interaction with both analyte and stationary phase implies the presence of secondary equilibration. Thus, the exploration of the MLC retention process is challenging [85, 87]. A QSRR-MLR modeling approach was performed by Ramezani et al. for testing anthraquinones. These authors linked molecular descriptors (partition coefficient calculated from hydrophobic fragmental constants (logP), Geary autocorrelation of lag 8 weighted by van der Waals volume descriptor (GATS8v), the mean topological charge index which represented the effect of analyte charge in the MLC separation (JGI4), and descriptors based on *3D molecule representation of structures based on electron diffraction theory* (3D-MoRSE), namely 3D-MoRSE descriptor of signal 27 (Mor27m) and 3D-MoRSE descriptor of Moran autocorrelation of lag 7 (MATS7md)) and empirical factors of six organic modifiers to anthraquinones' retention time. It was concluded that the retention behavior is significantly influenced by the modifier's logP values, as well as by the mass, molecular weight, and van der Waals volume, in addition to the topological charge [63].

Complementation of the available knowledge on MLC was attained by Krmar et al.; numerous mixed QSRR models were developed using different types of algorithms. Not only was the GBT identified as the most suitable but also the most significant properties relevant for the separation of aripiprazole and impurities were extracted. QSRR models, in addition to MDs, contained experimental parameter values (concentration of non-ionic surfactant Brij L23, pH, and the content of ACN) in line with the Box-Behnken design. Steric effects and dipole-dipole interactions were identified to be the most important thermodynamic molecular parameters relevant for retention behavior [27].

3.1.3 Cyclodextrin-modified liquid chromatography

Shifting the analytes' retention behavior in RP-HPLC can also be provoked by adding cyclodextrins (CD) to the mobile phase. Molecular retention patterns are modified because of CD-analyte complex formation, in addition to the adsorption process of CD on the stationary phase surface [85].

Maljurić et al. developed a QSRR-ANN model for the retention property analysis of risperidone, olanzapine, and related impurities in a CD-modified RP-HPLC system. The values of MSs, complex association constants, and chromatographic factors were used in the model. The most influential molecular descriptors and complex association constants were polarizability (POL), solvent-excluded volume (SEV), octanol/water partition coefficient (logP), dipole-dipole energy (DEN), binding energy (BE), electrostatic energy (EE), and unbound energy (UE) [48]. In a later study, a developed model was employed for determining a change in retention factor, the stability constants, and thermodynamic parameters of complex formation [88].

Another QSRR-ANN model for revealing separation processes in a CD-modified RP-HPLC system was developed by Đajić et al. The experimental parameters were acetonitrile percentage, aqueous phase pH, β-CD concentration, and column temperature. The most important molecular descriptors were identified as radial distribution function – 075/weighted by mass (RDF075m), signal 04/weighted by mass (Mor04v), and CATS2D positive-lipophilic at lag 08 (CATS2D_08_PL). It was found that the molecular size, shape, and lipophilicity of analytes significantly affect their retention. The retention behavior is also governed by the size and lipophilicity of the added CDs as it determines the structural agreement with the tested analytes [89].

3.2 QSRR for HPLC with unconventional stationary phases

Non-straightforward retention behavior resulting from the application of an unconventional stationary phase can be defined similarly as in the previous examples. As the QSRR successfully reveals additional interactions shaped by mobile phase modifiers, it can also expose multiple retention mechanisms provided by the stationary phase.

3.2.1 Immobilized artificial membrane chromatography

The characteristics of the stationary phase used in immobilized artificial membrane (IAM) chromatography are in line with its structure based on phosphatidylcholine residues covalently bound to silicon dioxide. In this way, the column mimics a phospholipid membrane monolayer and exhibits biomimetic properties [90].

In the research of Ciura et al., the general conclusions about the molecular retention mechanisms of isoxazolone on an IAM chromatographic system were derived from a QSRR model. The purpose of this research was to assess isoxazolone derivatives' affinity toward phospholipids. The model was developed using differential evolution combined with partial least squares regression (PLS). Molecular descriptors carrying the information referred to van der Waals volume as well as those defined based upon the weighted holistic invariant *molecular* theory (WHIM), geometry, topology, and atom-weights assembly theory (GETAWAY), and *3D molecule representation of structures based on electron diffraction theory* (3D MORSE), stood out as descriptors of importance, carrying the information related to molecular size, shape, symmetry, and atomic distribution. However, polarizability related and descriptors based on chemically advanced template search theory (CATS) were omitted despite being important for lipophilicity determination. The interpretation of these results led to a conclusion about the insufficient binding of isoxazolone derivatives to phospholipid molecules [90].

In another study, Buszevski et al. tried to gain insight into the biological activity of 30 flavonoids using IAM chromatographic analysis. The GA-PLS algorithm was used for QSRR model development. The conclusion about retention mechanisms was made upon quantum chemical descriptors, indicating that hydrophobic forces, dispersion effect, and electrostatic interactions govern the retention behavior of flavonoids in IAM chromatographic separation [36].

3.2.2 Mixed-mode liquid chromatography

A promising application of QSRR models has also been shown by the explanation of MMLC, where multiple functionalities in charge of providing different intermolecular interactions are integrated into a single stationary phase.

Obradović et al. developed QSRR models to characterize an MMLC system in which RP and hydrophilic interaction (HILIC) modes participate equally. Forty-three substances, serotonin, and imidazole receptor ligands were tested. Interestingly, separate QSRR models for four different types of responses were developed. The retention factor in pure eluents and the turning point for modality shifting were used as selected outputs. For characterizing the partition process in the RP mode, atomic mass, lipophilicity, and intermolecular hydrophobic interactions were proved to be important. The partition process in the HILIC modality was characterized by lipophilicity, distribution of ionic forms, and electrostatic properties. Adsorption, on the other hand, was driven by molecular geometry, electronegativity, polarizability, van der Waals volume, and atomic mass of the tested analytes. For the turning point and modality expressions, distribution of ionic forms, hydrogen bonding properties, and electronic properties, as well as atomic mass, were significant [91].

Russo et al. used a QSRR model developed by PLS in combination with block relevance (BR) to detect retention mechanisms provided by the arginine stationary phase. Due to the diverse interaction ability of the stationary phase, analytes with diverse ionization capacity (neutral, acids, and bases) were selected. It was noticed that the analyte's size and hydrogen donor capacity were important for the retention of neutral substances. For acidic molecules, descriptors calculated with VolSurf+ software and VS+ descriptors, did not describe the electric charge well enough; the MLR strategy was used for confirmation of the electrostatic background of acidic analytes' retention. Also, with the constructed QSRR model, it is possible to recognize the turning point for modality shifting. The basic substances did not show a sufficient degree of retention, so it was not possible to qualitatively define the retention mechanisms involved in their separation [92].

3.3 Future perspectives

With the use of adequate mathematical tools for linking input variables (both molecular descriptors and experimental parameters) with suitable responses, the statistical approach of QSRR modeling does not recognize limitations regarding the type of HPLC system that needs to be characterized. For this reason, it is considered that especially mixed QSRR models can significantly improve the understanding and development of HPLC methods when complex retention patterns are present due to a possible reduction of the requirements for experimental work.

4. Conclusion

It can be concluded from the literature that QSRR models have been widely applied in chromatographic science, this topic is, therefore, of great interest to researchers in different scientific areas. This chapter has presented the QSRR models with structuring possibilities in detail, the importance of molecular descriptors, and machine learning algorithms selection, as well as different approaches to conducting these important tasks. It can be also used as a guideline when choosing internal and external validation approaches to apply in the consideration of their main advantages and disadvantages. Special attention was put into disclosing the most important QSRR model applications, by pointing out the possibilities of investigating modified HPLC systems that are of great interest to analysts working with different kinds of compounds.

Acknowledgements

This work was supported by the Ministry of Education, Science and Technological Development, the Republic of Serbia through a grant agreement with the University of Belgrade – Faculty of Pharmacy No: 451-03-68/2022-2114/200161.

Conflict of interest

We declare that there is no conflict of interest.

Author details

Jovana Krmar, Bojana Svrkota, Nevena Đajić, Jevrem Stojanović, Ana Protić and Biljana Otašević*
Faculty of Pharmacy, University of Belgrade, Belgrade, Republic of Serbia

*Address all correspondence to: biljana.otasevic@pharmacy.bg.ac.rs

References

[1] Put R, Vander HY. Review on modelling aspects in reversed-phase liquid chromatographic quantitative structure–retention relationships. Analytica Chimica Acta. 2007; **602**(2):164-172. DOI: 10.1016/j.aca.2007.09.014

[2] Héberger K. Quantitative structure–(chromatographic) retention relationships. Journal of Chromatography. A. 2007;**1158**(1): 273-305. DOI: 10.1016/j.chroma.2007.03.108

[3] Kaliszan R. Chapter 17 - Quantitative structure property (Retention) relationships in liquid chromatography. In: Fanali S, Haddad PR, Poole CF, Schoenmakers P, Lloyd D, editors. Liquid Chromatography. Amsterdam: Elsevier; 2013. pp. 385-405. DOI: 10.1016/B978-0-12-415807-8.00017-1

[4] Muteki K, Morgado JE, Reid GL, Wang J, Xue G, Riley FW, et al. Quantitative structure retention relationship models in an analytical quality by design framework: Simultaneously accounting for compound properties, Mobile-phase conditions, and stationary-phase properties. Industrial and Engineering Chemistry Research. 2013; **52**(35):12269-12284. DOI: 10.1021/ie303459a

[5] Kaliszan R. QSRR: Quantitative structure-(chromatographic) retention relationships. Chemical Reviews. 2007; **107**(7):3212-3246. DOI: 10.1021/cr068412z

[6] Kaliszan R. Quantitative structure-retention relationships (QSRR) in chromatography. In: Wilson ID, editor. Encyclopedia of Separation Science. Oxford: Academic Press; 2000. pp. 4063-4075. DOI: 10.1016/b0-12-226770-2/01911-6

[7] Bodzioch K, Durand A, Kaliszan R, Bączek T, Vander HY. Advanced QSRR modeling of peptides behavior in RPLC. Talanta. 2010;**81**(4):1711-1718. DOI: 10.1016/j.talanta.2010.03.028

[8] Mauri A, Consonni V, Pavan M, Todeschini R. DRAGON software: An easy approach to molecular descriptor calculations. MATCH Communications in Mathematical and in Computer Chemistry. 2006;**56**: 237-248

[9] Golubović J. Application of Artificial Neural Networks in Building Models to Predict Retention Behaviour and Intensity of Mass Spectrometric Response in the Analysis of the Selected Azoles and Sartans by High Performance Liquid Chromatography. Belgrade: University of Belgrade - Faculty of Pharmacy; 2016

[10] Mao J, Akhtar J, Zhang X, Sun L, Guan S, Li X, et al. Comprehensive strategies of machine-learning-based quantitative structure-activity relationship models. iScience. 2021; **24**(9):103052-103052. DOI: 10.1016/j.isci.2021.103052

[11] Todeschini R, Consonni V. Handbook of Molecular Descriptors. Weinheim: Wiley VCHVerlag GmbH; 2000. p. 667. DOI: 10.1002/9783527613106

[12] Szucs R, Brown R, Brunelli C, Heaton JC, Hradski J. Structure driven prediction of chromatographic retention times: Applications to pharmaceutical analysis. International Journal of Molecular Sciences. 2021;**22**(8):3848

[13] Si-Hung L, Izumi Y, Nakao M, Takahashi M, Bamba T. Investigation of supercritical fluid chromatography retention behaviors using quantitative structure-retention relationships. Analytica Chimica Acta. 2022;**1197**: 339463-339463. DOI: 10.1016/j.aca.2022.339463

[14] Rojas C, Aranda JF, Pacheco Jaramillo E, Losilla I, Tripaldi P, Duchowicz PR, et al. Foodinformatic prediction of the retention time of pesticide residues detected in fruits and vegetables using UHPLC/ESI Q-orbitrap. Food Chemistry. 2021;**342**: 128354-128354. DOI: 10.1016/j.foodchem.2020.128354

[15] Park SH, Haddad PR, Talebi M, Tyteca E, Amos RIJ, Szucs R, et al. Retention prediction of low molecular weight anions in ion chromatography based on quantitative structure-retention relationships applied to the linear solvent strength model. Journal of Chromatography. A. 2017;**1486**:68-75

[16] D'Archivio AA, Maggi MA, Ruggieri F. Modelling of UPLC behaviour of acylcarnitines by quantitative structure–retention relationships. Journal of Pharmaceutical and Biomedical Analysis. 2014;**96**: 224-230. DOI: 10.1016/j.jpba.2014.04.006

[17] Akbar J, Iqbal S, Batool F, Karim A, Chan KW. Predicting retention times of naturally occurring phenolic compounds in reversed-phase liquid chromatography: A quantitative structure-retention relationship (QSRR) approach. International Journal of Molecular Sciences. 2012;**13**(11): 15387-15400. DOI: 10.3390/ijms131115387

[18] Oliveira TB, Gobbo-Neto L, Schmidt TJ, Da Costa FB. Study of chromatographic retention of natural terpenoids by chemoinformatic tools. Journal of Chemical Information and Modeling. 2015;**55**(1):26-38. DOI: 10.1021/ci500581q

[19] Dobričić V, Nikolic K, Vladimirov S, Čudina O. Biopartitioning micellar chromatography as a predictive tool for skin and corneal permeability of newly synthesized 17β-carboxamide steroids. European Journal of Pharmaceutical Sciences. 2014;**56**:105-112. DOI: 10.1016/j.ejps.2014.02.007

[20] Filipic S, Elek M, Nikolic K, Agbaba D. Quantitative structure-retention relationship Modeling of the retention behavior of guanidine and imidazoline derivatives in reversed-phase thin-layer chromatography. JPC Journal of Planar Chromatography Modern TLC. 2015;**28**(2):119-125. DOI: 10.1556/jpc.28.2015.2.6

[21] Karadžić Banjac M, Jevrić L, Kovačević S, Podunavac-Kuzmanovic S. Retention data from Normal-phase thin-layer chromatography in characterization of some 1,6-Anhydrohexose and D-Aldopentose derivatives by QSRR method. Journal of Liquid Chromatography and Related Technologies. 2015;**38**:1044-1044. DOI: 10.1080/10826076.2015.1012521

[22] Naylor BC, Catrow JL, Maschek JA, Cox JE. QSRR Automator: A tool for automating retention time prediction in Lipidomics and metabolomics. Metabolites. 2020;**10**(6):237

[23] Wang YT, Yang ZX, Piao ZH, Xu XJ, Yu JH, Zhang YH. Prediction of flavor and retention index for compounds in beer depending on molecular structure using a machine learning method. RSC Advances. 2021;**11**(58):36942-36950. DOI: 10.1039/D1RA06551C

[24] Amos RIJ, Haddad PR, Szucs R, Dolan JW, Pohl CA. Molecular modeling and prediction accuracy in quantitative structure-retention relationship calculations for chromatography. TrAC, Trends in Analytical Chemistry. 2018; **105**:352-359. DOI: 10.1016/j.trac.2018.05.019

[25] Bálint D, Jäntschi L. Comparison of molecular geometry optimization methods based on molecular descriptors. Mathematics. 2021;**9**(22):2855

[26] Amos RIJ, Tyteca E, Talebi M, Haddad PR, Szucs R, Dolan JW, et al. Benchmarking of computational methods for creation of retention models in quantitative structure–retention relationships studies. Journal of Chemical Information and Modeling. 2017;**57**(11):2754-2762. DOI: 10.1021/acs.jcim.7b00346

[27] Krmar J, Vukićević M, Kovačević A, Protić A, Zečević M, Otašević B. Performance comparison of nonlinear and linear regression algorithms coupled with different attribute selection methods for quantitative structure - retention relationships modelling in micellar liquid chromatography. Journal of Chromatography. A. 2020;**1623**: 461146-461146. DOI: 10.1016/j.chroma.2020.461146

[28] Talebi M, Schuster G, Shellie RA, Szucs R, Haddad PR. Performance comparison of partial least squares-related variable selection methods for quantitative structure retention relationships modelling of retention times in reversed-phase liquid chromatography. Journal of Chromatography. A. 2015;**1424**:69-76. DOI: 10.1016/j.chroma.2015.10.099

[29] González M, Terán C, Saíz-Urra L, Teijeira M. Variable selection methods in QSAR: An overview. Current Topics in Medicinal Chemistry. 2008;**8**:1606-1627. DOI: 10.2174/156802608786786552

[30] Urbanowicz RJ, Meeker M, La Cava W, Olson RS, Moore JH. Relief-based feature selection: Introduction and review. Journal of Biomedical Informatics. 2018;**85**:189-203. DOI: 10.1016/j.jbi.2018.07.014

[31] Goodarzi M, Jensen R, Vander HY. QSRR modeling for diverse drugs using different feature selection methods coupled with linear and nonlinear regressions. Journal of Chromatography B. 2012;**910**:84-94. DOI: 10.1016/j.jchromb.2012.01.012

[32] Hancock T, Put R, Coomans D, Vander Heyden Y, Everingham Y. A performance comparison of modern statistical techniques for molecular descriptor selection and retention prediction in chromatographic QSRR studies. Chemometrics and Intelligent Laboratory Systems. 2005; **76**(2):185-196. DOI: 10.1016/j.chemolab.2004.11.001

[33] Mizera M, Talaczyńska A, Zalewski P, Skibiński R, Cielecka-Piontek J. Prediction of HPLC retention times of tebipenempivoxyl and its degradation products in solid state by applying adaptive artificial neural network with recursive features elimination. Talanta. 2015;**137**:174-181. DOI: 10.1016/j.talanta.2015.01.032

[34] Daghir-Wojtkowiak E, Wiczling P, Bocian S, Kubik Ł, Kośliński P, Buszewski B, et al. Least absolute shrinkage and selection operator and dimensionality reduction techniques in quantitative structure retention relationship modeling of retention in hydrophilic interaction liquid chromatography. Journal of Chromatography. A. 2015;**1403**:54-62. DOI: 10.1016/j.chroma.2015.05.025

[35] Wen Y, Amos RIJ, Talebi M, Szucs R, Dolan JW, Pohl CA, et al. Retention index prediction using quantitative structure–retention relationships for improving structure identification in nontargeted metabolomics. Analytical Chemistry. 2018;**90**(15):9434-9440. DOI: 10.1021/acs.analchem.8b02084

[36] Buszewski B, Žuvela P, Sagandykova G, Walczak-Skierska J, Pomastowski P, David J, et al. Mechanistic chromatographic column characterization for the analysis of flavonoids using quantitative structure-retention relationships based on density functional theory. International Journal of Molecular Sciences. 2020;**21**(6):2053

[37] Park SH, De Pra M, Haddad PR, Grosse S, Pohl CA, Steiner F. Localised quantitative structure–retention relationship modelling for rapid method development in reversed-phase high performance liquid chromatography. Journal of Chromatography. A. 2020; **1609**:460508-460508. DOI: 10.1016/j.chroma.2019.460508

[38] Yan P, Wang L, Li S, Liu X, Sun Y, Tao J, et al. Improved structural annotation of triterpene metabolites of traditional Chinese medicine in vivo based on quantitative structure-retention relationships combined with characteristic ions: AlismatisRhizoma as an example. Journal of Chromatography B. 2021;**1187**: 123012-123012. DOI: 10.1016/j.jchromb.2021.123012

[39] Robnik-Šikonja M, Kononenko I. Theoretical and empirical analysis of ReliefF and RReliefF. Machine Learning. 2003;**53**(1):23-69. DOI: 10.1023/A:1025667309714

[40] Pawellek R, Krmar J, Leistner A, Djajić N, Otašević B, Protić A, et al. Charged aerosol detector response modeling for fatty acids based on experimental settings and molecular features: A machine learning approach. Journal of Cheminformatics. 2021;**13**(1): 53-53. DOI: 10.1186/s13321-021-00532-0

[41] Chen J, Tang YY, Fang B, Guo C. In silico prediction of toxic action mechanisms of phenols for imbalanced data with random Forest learner. Journal of Molecular Graphics & Modelling. 2012;**35**:21-27. DOI: 10.1016/j.jmgm.2012.01.002

[42] Gupta S, Basant N, Mohan D, Singh KP. Room-temperature and temperature-dependent QSRR modelling for predicting the nitrate radical reaction rate constants of organic chemicals using ensemble learning methods. SAR and QSAR in Environmental Research. 2016; **27**(7):539-558. DOI: 10.1080/1062936X.2016.1199592

[43] Agatonovic-Kustrin S, Beresford R. Basic concepts of artificial neural network (ANN) modeling and its application in pharmaceutical research. Journal of Pharmaceutical and Biomedical Analysis. 2000;**22**(5): 717-727. DOI: 10.1016/s0731-7085(99)00272-1

[44] Žuvela P, Macur K, Jay Liu J, Bączek T. Exploiting non-linear relationships between retention time and molecular structure of peptides originating from proteomes and comparing three multivariate approaches. Journal of Pharmaceutical and Biomedical Analysis. 2016;**127**: 94-100. DOI: 10.1016/j.jpba.2016.01.055

[45] Fatemi MH, Ghorbanzad'e M, Baher E. Quantitative structure retention relationship Modeling of retention time for some organic pollutants. Analytical Letters. 2010;**43**(5):823-835. DOI: 10.1080/00032710903486294

[46] Mozafari Z, Arab Chamjangali M, Arashi M, Goudarzi N. QSRR models for predicting the retention indices of VOCs in different datasets using an efficient variable selection method coupled with artificial neural network modeling: ANN-based QSPR modeling. Journal of the Iranian Chemical Society. 2022; **19**(6):2617-2630. DOI: 10.1007/s13738-021-02488-2

[47] Noorizadeh H, Farmany A, Narimani H, Noorizadeh M. QSRR using evolved artificial neural network for 52 common pharmaceuticals and drugs of abuse in hair from UPLC–TOF-MS. Drug Testing and Analysis. 2013;**5**(5):320-324. DOI: 10.1002/dta.309

[48] Djajić N, Golubović J, Otašević B, Zecevic M, Protić A. Quantitative structure –retention relationship modeling of selected antipsychotics and their impurities in green liquid chromatography using cyclodextrin mobile phases. Analytical and Bioanalytical Chemistry. 2018;**410**: 2533-2550. DOI: 10.1007/s00216-018-0911-3

[49] Golubović J, Protić A, Otašević B, Zečević M. Quantitative structure–retention relationships applied to development of liquid chromatography gradient-elution method for the separation of sartans. Talanta. 2016;**150**: 190-197. DOI: 10.1016/j.talanta.2015.12.035

[50] D'Archivio A, Maggi M, Ruggieri F. Artificial neural network prediction of multilinear gradient retention in reversed-phase HPLC: Comprehensive QSRR-based models combining categorical or structural solute descriptors and gradient profile parameters. Analytical and Bioanalytical Chemistry. 2014;**407**: 1181-1190. DOI: 10.1007/s00216-014-8317-3

[51] Dobričić V, Savić J, Nikolic K, Vladimirov S, Vujić Z, Brborić J. Application of biopartitioning micellar chromatography and QSRR modeling for prediction of gastrointestinal absorption and design of novel β-hydroxy-β-arylalkanoic acids. European Journal of Pharmaceutical Sciences. 2017;**100**: 280-284. DOI: 10.1016/j.ejps.2017.01.023

[52] Parinet J. Predicting reversed-phase liquid chromatographic retention times of pesticides by deep neural networks. Heliyon. 2021;**7**(12):e08563-e08563. DOI: 10.1016/j.heliyon.2021.e08563

[53] Ju R, Liu X, Zheng F, Lu X, Xu G, Lin X. Deep neural network pretrained by weighted autoencoders and transfer learning for retention time prediction of small molecules. Analytical Chemistry. 2021;**93**(47):15651-15658. DOI: 10.1021/acs.analchem.1c03250

[54] Pasin D, Mollerup CB, Rasmussen BS, Linnet K, Dalsgaard PW. Development of a single retention time prediction model integrating multiple liquid chromatography systems: Application to new psychoactive substances. Analytica Chimica Acta. 2021;**1184**:339035-339035. DOI: 10.1016/j.aca.2021.339035

[55] Randazzo GM, Bileck A, Danani A, Vogt B, Groessl M. Steroid identification via deep learning retention time predictions and two-dimensional gas chromatography-high resolution mass spectrometry. Journal of Chromatography. A. 2020;**1612**: 460661-460661. DOI: 10.1016/j.chroma.2019.460661

[56] Chen H, Engkvist O, Wang Y, Olivecrona M, Blaschke T. The rise of deep learning in drug discovery. Drug Discovery Today. 2018;**23**(6):1241-1250. DOI: 10.1016/j.drudis.2018.01.039

[57] Adadi A. A survey on data-efficient algorithms in big data era. Journal of Big Data. 2021;**8**(1):24-24. DOI: 10.1186/s40537-021-00419-9

[58] Ciura K, Pastewska M, Ulenberg S, Kapica H, Kawczak P, Bączek T. Chemometric analysis of bio-inspired micellar electrokinetic chromatographic systems – Modelling of retention mechanism and prediction of biological properties using bile salts surfactants. Microchemical Journal. 2021;**167**:106340-106340. DOI: 10.1016/j.microc.2021.106340

[59] Aalizadeh R, Nika MC, Thomaidis N. Development and application of retention time prediction models in the suspect and non-target screening of emerging contaminants. Journal of Hazardous Materials. 2019;**363**:275-288. DOI: 10.1016/j.jhazmat.2018.09.047

[60] Riahi S, Pourbasheer E, Ganjali MR, Norouzi P. Investigation of different linear and nonlinear chemometric methods for modeling of retention index of essential oil components: Concerns to support vector machine. Journal of Hazardous Materials. 2009;**166**(2):853-859. DOI: 10.1016/j.jhazmat.2008.11.097

[61] Zhang X, Zhang X, Li Q, Sun Z, Song L, Sun T. Support vector machine applied to study on quantitative structure–retention relationships of polybrominated diphenyl ether congeners. Chromatographia. 2014;**77**:1387-1398. DOI: 10.1007/s10337-014-2735-4

[62] Song M, Breneman CM, Bi J, Sukumar N, Bennett KP, Cramer S, et al. Prediction of protein retention times in anion-exchange chromatography systems using support vector regression. Journal of Chemical Information and Computer Sciences. 2002;**42**(6):1347-1357. DOI: 10.1021/ci025580t

[63] Čolović J, Kalinić M, Vemić A, Eric S, Malenović A. Investigation into the phenomena affecting the retention behavior of basic analytes in chaotropic chromatography: Joint effects of the most relevant chromatographic factors and analytes' molecular properties. Journal of Chromatography. A. 2015;**1425**:150-157. DOI: 10.1016/j.chroma.2015.11.027

[64] Svetnik V, Wang T, Tong C, Liaw A, Sheridan RP, Song Q. Boosting: An ensemble learning tool for compound classification and QSAR Modeling. Journal of Chemical Information and Modeling. 2005;**45**(3):786-799. DOI: 10.1021/ci0500379

[65] Breiman L, Friedman JH, Olshen RA, Stone CJ. Classification and Regression Trees. 1st ed. New York: Routledge; 1984. p. 368. DOI: 10.1201/9781315139470

[66] Goudarzi N, Shahsavani D, Emadi-Gandaghi F, Chamjangali MA. Application of random forests method to predict the retention indices of some polycyclic aromatic hydrocarbons. Journal of Chromatography. A. 2014;**1333**:25-31. DOI: 10.1016/j.chroma.2014.01.048

[67] Svetnik V, Liaw A, Tong C, Culberson JC, Sheridan RP, Feuston BP. Random Forest: A classification and regression tool for compound classification and QSAR Modeling. Journal of Chemical Information and Computer Sciences. 2003;**43**(6):1947-1958. DOI: 10.1021/ci034160g

[68] Wang C, Skibic MJ, Higgs RE, Watson IA, Bui H, Wang J, et al. Evaluating the performances of quantitative structure-retention

relationship models with different sets of molecular descriptors and databases for high-performance liquid chromatography predictions. Journal of Chromatography. A. 2009;**1216**(25): 5030-5038. DOI: 10.1016/j.chroma.2009.04.064

[69] Yang JJ, Han Y, Mah CH, Wanjaya E, Peng B, Xu TF, et al. Streamlined MRM method transfer between instruments assisted with HRMS matching and retention-time prediction. Analytica Chimica Acta. 2020;**1100**:88-96. DOI: 10.1016/j.aca.2019.12.002

[70] Goudarzi N, Shahsavani D. Application of a random forests (RF) method as a new approach for variable selection and modelling in a QSRR study to predict the relative retention time of some polybrominateddiphenylethers (PBDEs). Analytical Methods. 2012;**4**: 3733-3738. DOI: 10.1039/c2ay25484k

[71] Hastie T, Tibshirani R, Friedman J. Boosting and additive trees. In: Hastie T, Tibshirani R, Friedman J, editors. The Elements of Statistical Learning: Data Mining, Inference, and Prediction. New York: Springer New York; 2009. pp. 337-387. DOI: 10.1007/978-0-387-84858-7_10

[72] Bauer E, Kohavi R. An empirical comparison of voting classification algorithms: Bagging, boosting, and variants. Machine Learning. 1999;**36**(1): 105-139. DOI: 10.1023/A:1007515423169

[73] Bouwmeester R, Martens L, Degroeve S. Comprehensive and empirical evaluation of machine learning algorithms for small molecule LC retention time prediction. Analytical Chemistry. 2019;**91**(5):3694-3703. DOI: 10.1021/acs.analchem.8b05820

[74] Liapikos T, Zisi C, Kodra D, Kademoglou K, Diamantidou D, Begou O, et al. Quantitative structure retention relationship (QSRR) modelling for analytes' retention prediction in LC-HRMS by applying different machine learning algorithms and evaluating their performance. Journal of Chromatography B. 2022;**1191**: 123132-123132. DOI: 10.1016/j.jchromb.2022.123132

[75] Veerasamy R, Rajak H, Jain A, Sivadasan S, Christapher PV, Agrawal R. Validation of QSAR models - strategies and importance. International Journal of Drug Design and Discovery. 2011;**2**: 511-519

[76] Roy K, Mitra I, Kar S, Ojha PK, Das RN, Kabir H. Comparative studies on some metrics for external validation of QSPR models. Journal of Chemical Information and Modeling. 2012;**52**(2): 396-408. DOI: 10.1021/ci200520g

[77] Kiralj R, Ferreira M. Basic validation procedures for regression models in QSAR and QSPR studies: Theory and application. Journal of the Brazilian Chemical Society. 2008;**20**:770-787. DOI: 10.1590/S0103-50532009000400021

[78] Tropsha A. Best practices for QSAR model development, validation, and exploitation. Molecular Informatics. 2010;**29**(6–7):476-488. DOI: 10.1002/minf.201000061

[79] Parinet J. Prediction of pesticide retention time in reversed-phase liquid chromatography using quantitative-structure retention relationship models: A comparative study of seven molecular descriptors datasets. Chemosphere. 2021;**275**:130036-130036. DOI: 10.1016/j.chemosphere.2021.130036

[80] Roy K, Ambure P, Aher RB. How important is to detect systematic error in predictions and understand statistical applicability domain of QSAR models?

Chemometrics and Intelligent Laboratory Systems. 2017;**162**:44-54. DOI: 10.1016/j.chemolab.2017.01.010

[81] Taraji M, Haddad PR, Amos RIJ, Talebi M, Szucs R, Dolan JW, et al. Error measures in quantitative structure-retention relationships studies. Journal of Chromatography. A. 2017;**1524**: 298-302. DOI: 10.1016/j.chroma.2017.09.050

[82] Roy K, Kar S, Ambure P. On a simple approach for determining applicability domain of QSAR models. Chemometrics and Intelligent Laboratory Systems. 2015;**145**:22-29. DOI: 10.1016/j.chemolab.2015.04.013

[83] Roy K, Kar S, Das RN. Statistical methods in QSAR/QSPR. In: Roy K, Kar S, Das RN, editors. A Primer on QSAR/QSPR Modeling: Fundamental Concepts. Cham: Springer International Publishing; 2015. pp. 37-59. DOI: 10.1007/978-3-319-17281-1_2

[84] Haddad PR, Taraji M, Szücs R. Prediction of analyte retention time in liquid chromatography. Analytical Chemistry. 2021;**93**(1):228-256. DOI: 10.1021/acs.analchem.0c04190

[85] Djajić N, Krmar J, Rmandić M, Rašević M, Otašević B, Zečević M, et al. Modified aqueous mobile phases: A way to improve retention behavior of active pharmaceutical compounds and their impurities in liquid chromatography. Journal of Chromatography Open. 2022; **2**:100023-100023. DOI: 10.1016/j.jcoa.2021.100023

[86] Cecchi T, Passamonti P. Retention mechanism for ion-pair chromatography with chaotropic reagents. Journal of Chromatography. A. 2009;**1216**(10): 1789-1797. DOI: 10.1016/j.chroma.2008.10.031

[87] Ramezani A, Yousefinejad S, Shahsavar A, Mohajeri A, Absalan G. Quantitative structure-retention relationship for chromatographic behaviour of anthraquinone derivatives through considering organic modifier features in micellar liquid chromatography. Journal of Chromatography A. 2019;**1599**:46-54. DOI: 10.1016/j.chroma.2019.03.063

[88] Djajić N, Otašević B, Malenović A, Zecevic M, Protić A. Quantitative structure retention relationship modeling as potential tool in chromatographic determination of stability constants and thermodynamic parameters of β-cyclodextrin complexation process. Journal of Chromatography. A. 2020;**1619**: 460971-460971. DOI: 10.1016/j.chroma.2020.460971

[89] Djajić N, Petković M, Zečević M, Otašević B, Malenović A, Holzgrabe U, et al. A comprehensive study on retention of selected model substances in β-cyclodextrin-modified high performance liquid chromatography. Journal of Chromatography. A. 2021; **1645**:462120-462120. DOI: 10.1016/j.chroma.2021.462120

[90] Ciura K, Fedorowicz J, Žuvela P, Lovrić M, Kapica H, Baranowski P, et al. Affinity of antifungal Isoxazolo[3,4-b] pyridine-3(1H)-ones to phospholipids in immobilized artificial membrane (IAM) chromatography. Molecules. 2020;**25**(20):4835

[91] Obradović D, Oljačić S, Nikolić K, Agbaba D. Investigation and prediction of retention characteristics of imidazoline and serotonin receptor ligands and their related compounds on mixed-mode stationary phase. Journal of Chromatography. A. 2019;**1585**:92-104. DOI: 10.1016/j.chroma.2018.11.051

[92] Russo G, Vallaro M, Cappelli L, Anderson S, Ermondi G, Caron G. Characterization of the new CelerisTM arginine column: Retentive behaviour through a combination of chemometric tools and potential in drug analysis. Journal of Chromatography. A. 2021; **1651**:462316-462316. DOI: 10.1016/j.chroma.2021.462316